# Installation Effects in Fan Systems

# Conference Planning Panel

W T W Cory, CEng, MIMechE, MCIBSE, MIAgrE (Chairman)
*Woods of Colchester Limited*
*Colchester*

C W Lack, BSc
*Elta Fans Limited*
*Byfleet*

A N Bolton, BSc
*National Engineering Laboratory*
*East Kilbride*

R L Elder BSc, PhD
*School of Mechanical Engineering*
*Cranfield Institute of Technology*
*Bedford*

D Koblenz
*Ziehl Abegg GmbH*
*Kunzelsau*
*West Germany*

S Becirspahic
*CETIAT*
*Orsay*
*France*

# Proceedings of the Institution of Mechanical Engineers

## European Conference

## Installation Effects in Fan Systems

14–15 March 1990
The Institution of Mechanical Engineers
Birdcage Walk
London

Sponsored by
Power Industries Division of the Institution of Mechanical Engineers

IMechE 1990–1

 Published for the Institution of Mechanical Engineers by
Mechanical Engineering Publications Limited

The Publishers are not responsible for any statement made in this publication. Data, discussion and conclusions developed by authors are for information only and are not intended for use without independent substantiating investigation on the part of potential users.

Printed by Waveney Print Services Ltd, Beccles, Suffolk

# Contents

# Installation effects in fan systems

**A N BOLTON**, BSc, **A J GRAY** and **E J MARGETTS**, MIOA
National Engineering Laboratory, East Kilbride, Glasgow

SYNOPSIS

The paper reviews some of the changes in fan aerodynamic, acoustic and vibration performance which can be attributed to interactions between a fan and the system in which it operates. The need for the recognition of the importance of installation effects by manufacturers, specifiers and users of fan systems is highlighted by reference to commercial, efficiency and safety aspects of fan systems. It is concluded that there is an urgent requirement for further research.

## 1   INTRODUCTION

It is difficult to give a succinct definition of the term 'Fan Installation Effect'. At its broadest, an Installation Effect can be defined as being the cause of some aspect of the performance of a fan in a system being different from that which would be anticipated from test data available on the performance of the fan in an ideal test configuration.

By way of example we can consider the aerodynamic performance of a fan in a system. In a standardised airway test at a manufacturer's works the fan aerodynamic performance will be determined over a range of flowrates so that at flowrate, Q, the static pressure rise is $P_f$. In a standardised airway test a component manufacturer will determine the pressure loss of a component as a function of flowrate so that at flowrate, Q, the pressure loss is $P_c$. It could thus be anticipated that, if the fan and component are connected together then, at flowrate, Q, the effective pressure rise would be $(P_f - P_c)$. If the measured value of pressure rise were $P_s$, then, if $P_s$ was not equal to $(P_f - P_c)$, it would be said that the fan performance has been affected by the installation of the component. The magnitude of the difference between $(P_f - P_c)$ and $P_s$ would be the magnitude of the installation effect.

Installation effects can affect not only the aerodynamic, acoustic and vibration performance of a fan but also that of static ductwork fittings such as bends, branches, diffusers and attenuators.

Identification of the existence of an installation effect raises several general questions. Why are installation effects now a topic of interest? Are they of significant importance? To whom are they of interest?

A number of factors can be identified as answers to the first question.

● There is now more testing of the individual components comprising a system, a move being increasingly stimulated by Quality Assurance requirements and a demand for product performance certification. Testing is now generally more comprehensive through the use of automated testing and logging equipment and accurate data on the performance of individual components in idealised test arrangements is more readily available.

● There is now a more critical appraisal of the performance of systems at the commissioning stage. In part, this derives from the fact that systems are becoming increasingly complex and require to be operated under accurately known and closely controlled conditions. Also, commissioning engineers have developed more sophisticated measurement techniques and can determine, with a fair degree of accuracy, component performance within a system.

Comparison of the measured system performance with the predicted performance will yield information on any installation effects, though this is not usually done unless a major problem has arisen with the system performance.

● Advances in design and manufacturing technology have resulted in fans which, for a given duty, are more efficient, smaller, of lighter weight construction and of higher specific power than previously. There is evidence to show that the performance of such machines is more susceptible to installation effects. In the same way, ductwork has become of lighter weight construction with less rigid supports. Any instabilities in flow or increases in noise are thus potentially more noticeable.

● Changes in the products being offered by manufacturers have also had a significant influence on the awareness of installation effects on fan performance. The last 10 years has seen the creation of a new market in the supply of Air Handling Units, where one manufacturer has the responsibility for ensuring the aerodynamic, acoustic, thermal and vibration performance of a unit incorporating fans, heaters, coolers, humidifiers, filters and silencers. Comparison of the actual unit performance with the sum of the individual component performances has highlighted major differences.

Whilst the above factors are of technical interest, the ultimate importance of understanding and quantifying installation effects is commercial. With a dearth of reliable, quantitative information on installation effects being available, there are different ways in which the installation effect problem may be tackled at project design and tendering stages. Two extreme approaches can be identified. In the first, it is assumed that any installation effects are negligible and components are selected on the basis of just being able to meet the specification. This approach will lead to the least costly system. However, if problems do arise because of installation effects they can be difficult and expensive to rectify. It may not be possible to increase fan speed, for example, because of electric motor power limitations or, worse still, because of power supply limitations. Space restrictions may limit the ability to install alternative, larger units. This approach is essentially a high risk route.

An alternative approach would be to allow a generous margin for the uncertainties of the performance of the individual system components. This approach is likely to lead to an expensive system and unlikely to lead to a contract! The application of safety margins can itself lead to operating problems. For example, the specifier of a system may incorporate an initial margin on flowrate of 10 per cent to allow for a possible future expansion. This means, in effect, a 20 per cent increase in pressure. The specifier of the system may build in a 10 per cent extra pressure allowance in specifying the fan duty to cover possible uncertainty in system loss calculations. Finally, the fan supplier may add a 5 per cent margin to ensure that his fan will meet the specification. The net effect is that the fan is considerably oversized, will operate away from its best efficiency and lowest noise duty and will have a reduced stall margin. The capital cost is increased and running costs are significantly higher. The use of some form of variable blade pitch, guide vane setting or speed control on a fan can allow for some margin of uncertainty in estimating system resistance.

Most systems require to be balanced during commissioning. This is most commonly achieved by introducing some form of resistive losses which is inherently uneconomic in terms of operating costs. It has been estimated (1) that the electrical power consumed by fans within the EEC is 30 000 000 MWh per annum. In many systems fans are unlikely to be perfectly matched to the duty requirements and will be operating away from their peak efficiency duties or generating excess pressure which has to be subsequently dissipated. It is clear that a very modest improvement in operating fans closer to their peak efficiency duty by better matching to the system would offer huge energy savings.

One of the questions posed at the start of the paper was "Who is responsible and who should be interested in installation effects"? The fan manufacturer can demonstrate the performance of his fan in a standard test rig and 'prove' it performs satisfactorily. A system component manufacturer can test his component, say a bend, and show it has a certain pressure loss and acoustic performance. Who is then responsible for the fact that, in combination, the resulting performance is not the sum of the two individual parts? Is it either of the manufacturers? Is it the specifier, the installer or the user? Clearly, all parties have an interest; but it requires a Solomon to apportion responsibility.

This paper surveys the broad scope of fan installation effects indicating aspects of aerodynamic, acoustic and vibration performance where effects are known to be significant. The basis for assessing the magnitude of any installation effect on fan performance is a comparison between the performance determined in a standardised test installation and that measured in situ. In the UK the relevant fan performance standards are the various parts of BS 848, (2) which look increasingly likely to form the basis of future ISO (3) and CEN standards for fans.

## 2    AERODYNAMIC INSTALLATION EFFECTS

### 2.1    Determination of Standardized Aerodynamic Performance Data

In the UK, the first direct move to recognise the influence of fan installation on standardised performance was made about 20 years ago when the revision of the British Standard BS 848 was being considered. This standard now specifies four different installation configurations, viz

type A – free inlet    : free outlet,

type B – free inlet    : ducted outlet,

type C – ducted inlet : free outlet, and

type D – ducted inlet : ducted outlet,

recognizing that the aerodynamic performance of a fan could be strongly influenced by the presence of connecting ductwork.

It should be noted that there is no such concept as the 'absolute performance' of a fan. The definitions of performance in the British or International Standard are, in essence, conventional values and really only relate to the performance obtained under the specified duct installation conditions. In the case of tests in installation configurations B and D, the outlet pressure is measured five diameters downstream of the fan after a flow straightener section two diameters long. The static pressure at the fan outlet flange is calculated by adding friction losses, calculated according to a prescribed formula, to the measured value.

Inclusion of the straightener specified by the test standard is to ensure the removal of the swirl component which has an influence on the pressure distribution across the duct and prevents accurate determination of the static pressure energy from measurements taken at wall tappings. Irrespective of the amount of energy actually lost in the straightener, the loss of static pressure through the straightener which can be credited to the fan is limited to a defined and conventional value. For fans with a high proportion of swirl in the outlet flow the static performance as determined in a standardised test rig may result in a considerably reduced static pressure compared with that which could be obtained in some other test arrangement.

It was found from extensive testing that the five diameter length of duct from fan to measurement plane was a 'reasonable practice'. Various papers have presented different distances at which the wall static pressure achieves its fully developed value. The American Air Moving and Control Association (AMCA), has published a document AMCA 201, (4) gives a rule-of-thumb such that full pressure is achieved at a distance of (2.5 + V/5) diameters, where V is the axial velocity in m/s. The document presents a table giving reductions in pressure as a function of having shorter lengths of straight duct. Deeprose (5) indicated that, for a non-guide vane axial fan, static pressure could be maximised by having a six diameter long discharge duct without a flow straightener. Cory (6) has reported that the standardised length of discharge duct is insufficient to allow complete diffusion of high velocity from centrifugal impellers which are generally offset from the centre line of any duct and thus results in an underestimate of the potential aerodynamic performance.

The four installation categories, A, B, C and D represent practical limits on the range of possible standardised installations for which fan performance could be sensibly compared and they cover the cases of the fan having either no connected ducting so that the flow is drawn from or discharged to an effectively infinite space or having straight lengths of duct, in effect, sufficiently long to ensure fully developed flow conditions. In general, it has been found that the aerodynamic performance achieved in installation types A and C is similar, whilst the performance of fans in installation configurations B and D is also similar.

## 2.2 Examples of various installation conditions

In most fan systems the inlet and outlet connections are likely to depart from the idealised standard conditions. Duct components such as bends, silencers or changes of section are often located close to the fan. These components can alter the profile and distribution of the flow into or out of the fan and it is the distorted flow which affects the performance of the fan. Several examples of different installations are shown in Fig. 1.

## 2.3 Predicting the installation effect

Mathematical modelling of installation effects on a fan is extremely challenging and, at present, design engineers have to rely on past experience or on model testing of proposed systems. A series of data published by AMCA (4) provides the most comprehensive attempt yet to quantify aerodynamic installation effects. The data comprises a chart, see Fig. 2a, on which are plotted a set of lines representing static pressure loss as a function of the square of flowrate. These lines are called System Effect Curves. Each line is identified by a letter, the greatest system effect being represented by the line A. A series of tables, as shown for example in Fig. 2b, then indicate the system effect curve appropriate to a particular fan and component installation configuration. The predicted performance of the fan plus component can then be calculated from the relationship

Estimated = Fan − component + system effect
pressure     pressure   loss        factor

The concept of the system effect factor is a useful one and the data is based on a considerable volume of test data. However, the data does only cover a limited number of installation conditions. As further information is gathered the correlations are modified. Recently, after a series of tests (7) on one vane axial fan and one tube axial fan, it has been recommended that the system effects applicable to certain axial fan and right-angled bend installation configuration should be amended. In some cases the installation effect changed by eight lines from line 0 to line W. Such a large change indicates that there can be problems in trying to assess system effect data and some of these are outlined below. It may suggest that the original data was of doubtful validity or that the system effect factor is more closely related to the actual design of the fan tested rather than to the generic type of which it is representative. If the latter case is true, then it will be very difficult for system designers to produce accurate estimates of system performance without resorting to physical testing.

Apparent shifts in fan performance need not always indicate that the system is influencing the fan's performance. Manufacturing tolerances can be such that there can be significant changes between the performances of nominally identical fans. Changes in tip clearance on axial fans (8) and changes to the clearance between a centrifugal impellers and the inlet eye (9, 10) can have significant effects on the aerodynamic performance as can differences in blade setting angles, blade roughness etc.

## 2.4 Problems in quantifying system effects

### 2.4.1 Pressure loss of single static components

Values for the static pressure loss of single components with no moving parts such as bends can be found in many reference books. Different books quote different figures. It may be rather cynical but if two values agree, then it is more likely that they are from the same original source than that they represent corroborating independent assessments. One problem is that there is no recognised test method for determining the pressure loss of a component. Although the flow into a duct component can be fully developed, the outlet flow almost certainly have a distorted profile and may contain swirl. The pressure at wall tappings is likely to vary with distance from the outlet of the component. What value should be taken to specify the pressure loss? What allowance should be made for duct friction between the discharge of the component and the measurement plane? If different criteria are used then different losses will be measured for the same component.

In a recent series of experiments partly funded by the EEC, four laboratories - CETIAT in France, University of Louvian in Belgium, University of Mannheim in Germany and NEL in the

UK — measured the static pressure losses of two components. Each laboratory used the same design of test rig shown in Fig. 3a. The essential features of the rig are the perforated inlet plate and the long inlet duct to ensure developed flow at the inlet of the component and the outlet duct incorporating an etoile flow straightener which is the same as the 'outlet part' defined in Part 1 of BS 848. Pressures are measured using wall static tappings at planes 5D and 5.5D upstream and downstream of the component. The friction loss and the loss through the etoile are as specified in BS 848 : Part 1. The measured pressure loss for one of the components is shown in Fig. 3b. As is the case for fans, the pressure loss is not an absolute quantity but is a 'conventional value' defined with reference to the standardised installation.

As Fig. 3b shows, the pressure loss is a function of Reynolds number. The pressure loss is also very dependant on the geometry and manufacturing finish of the individual component. Tests undertaken at CETIAT (11) on five commercially-available, 250 mm diameter, segmented bends made by different manufacturers, showed a spread of loss coefficients from 0.30 to 0.75. Similar components tested at 400 mm diameter had a range of loss coefficients of 0.3 to 0.5. Such a wide scatter of data highlights the problems which face the design engineer in trying to assess the resistance of even a simple ductwork system.

## 2.4.2 Pressure loss of multiple static components

Two static components placed close to each other produce an 'installation effect'. For example, the pressure loss of two right-angled bends is not twice the pressure loss of a single bend but, depending on the configuration and spacing between bends, may be more or less. Data on combined losses is less freely available than for single components. Miller (12) suggests that two closely-coupled bends will generally have a lower pressure loss than two single bends and provides comprehensive tables giving loss corrections as functions of the relative orientation of the bends and their separation. For virtually all cases the combined loss is lower than two single losses. Jorgensen (13) shows both increases and decreases, with the magnitude of the increases significantly greater than quoted by Miller. AMCA publication 200 (14) indicates that two bends will have a greater loss and quotes a loss coefficient of 4.2 for two bends separated by a 1D straight section compared with 2.4 for the sum of two individual bend losses. The equivalent loss coefficient suggested by Miller is 1.9.

AMCA 200 indicates that if components are separated by more than 10D the installation effect may be ignored. Miller cautions that with two right-angled bends configured to generate swirling flow, appreciable swirl will exist up to 100 diameters downstream of the second bend.

## 2.4.3 Research into installation effects

### Ductwork components

Recognising the lack of available published data, the Community Bureau of Reference (BCR) of the EEC is sponsoring research at the four laboratories mentioned in Section 2.4.1 to determine pressure losses for a range of different duct fittings including bends, diffusers, junctions, contractions and expansions. The data from the tests will be used to form a reference data bank of losses for both isolated and combined components which can be accessed by designers.

### Fans

Various manufacturers and system designers have built up sets of installation effect factors as a result of past experience which are usually for exclusive, in-house use. AMCA have taken the lead in publishing information and have a programme of regularly revising and updating it. In the UK, the Fan Manufacturers Association (FMA) are sponsoring investigations at NEL into the effect of closely-connected ductwork fittings on the aerodynamic performance of eight different fans covering a range of axial, mixed and radial flow types. The initial programme is scheduled to last 18 months.

## 3  ACOUSTIC INSTALLATION EFFECTS

### 3.1  Determination of standardised acoustic performance data

BS 848 : Part 2 : 1985 will form the basis for determining the magnitude of any acoustic installation effects. Like the aerodynamic performance standard, the acoustic standard recognises the four installation categories A, B, C and D. However, the acoustic situation is more complex because inlet noise, outlet noise and casing radiated noise require to be seperately identified for each installation configuration. Additionally, the accuracy with which noise can be assessed, particularly under site conditions, means that it may not always be possible to distinguish installation effects from measurement uncertainty.

There are two aspects which have to be considered in relation to noise installation effects. These are the noise generation mechanisms and the noise radiation or transmission mechanisms and each is, to some extent, dependant on the other.

### Noise Generation Mechanisms

Noise in most fans is caused by fluctuating aerodynamic forces on the blades (15, 16) although there are contributions from the displacement of air by the blades and from the turbulence generated in the airstream. Fluctuating forces on the blade surfaces arise from turbulence in the boundary layer causing local pressure fluctuations at the blade surface and from more extended changes in the pressure distribution over the blade surfaces which arise from the interaction of the blade with fluctuating incident flow and incoming turbulence. Different installation conditions can affect the blade boundary layers only to a limited extent. The main influences on noise generation are incident turbulence and asymmetries in the inlet flow.

Readily identifiable sources of turbulence are duct wall boundary layers, non-streamlined inlets where the fan blade tips are operating in highly turbulent flows and objects upstream of the fan causing wakes. Most handbooks on noise feature examples of poor installations which generate excess turbulence in the inflow.

The response of an aerofoil blade to turbulence has been studied by a number of researchers (17, 18). Their analyses showed that the response is governed by a number of features of the aerofoil shape and the nature of the unsteady inlet flow. The change in blade lift force due to the gust may be described by a relationship of the type

$$\frac{\text{fluctuating lift}}{\text{steady lift}} = \frac{\text{fluctuating velocity}}{\text{steady velocity}} \times \frac{\text{response}}{\text{function}}$$

There are a number of different response functions but they can be typified by the response of an isolated aerofoil to a transverse gust, ie a perturbation in velocity perpendicular to the blade chord. For this case, the response function is known as the Sears function, $S(\omega)$, which has the form shown in Fig. 4.

$\omega$ is a non-dimensional frequency which can be most easily considered in terms of the ratio between the wavelength of the disturbance and the blade chord so that

$$\omega = \frac{2\pi(\text{blade chord})}{\text{wavelength of inlet turbulence}}$$

or
$$= \frac{2\pi c}{\lambda}$$

$$= \frac{\nu c}{2U}$$

where U is the longitudinal velocity of the flow over the blade, and

$\nu$ is the frequency of the turbulent gusts.

It is important to note that the frequency of the turbulence is defined in a frame of reference relative to the blade. Thus in Fig. 5, which shows elongated turbulent eddies being convected onto a blade, the frequency spectrum which would be recorded by a hot wire positioned upstream of the blade would show dominant low frequency components. The blade would, however, experience a perturbation lasting only for the short time it takes for the blade to traverse the width of the eddy and this represents a much higher frequency.

It can be seen from Fig. 4 that the response is greatest for small values of $\omega$ which implies that small scale, high frequency turbulence will generate less change in lift than larger scale variations.

Sharland (19) derived an expression relating the radiated sound power generated by a fluctuating blade loads. It took the form,

$$W \propto \left|\frac{dL}{dt}\right| l_s l_c$$

where $l_s l_c$ defines the area of blade surface over which the fluctuating force is acting. The important point is that the sound power is proportional to the square of the magnitude of the fluctuating lift. In some way, therefore, a fan can be regarded as an amplifying transformer, transforming turbulence in the air flow into noise (20).

The flow field into a fan can be characterised as being steady, steady with spatially non-uniform flow, or unsteady. In practical fan installations a uniformly steady flow field will not exist in an ordinary fan installation. A

steady flow field with spatially non-uniform flow would be characterised by flow which was essentially constant in time but with different velocities at different points. Examples of this would be the flow downstream of a set of vanes or obstructions, typified by the presence of wakes, or the flow downstream of a bend where there would be an asymmetric flow profile. As the blades pass through the disturbance they experience a fluctuation in lift and hence radiate a pressure variation. Since the position of the distorted flow is invariant, the regular interception of the flow disturbance by the blades generates a regular pressure disturbance which is perceived as discrete tones at the shaft and blade passing frequencies and their higher harmonics. In practice, the wakes fluctuate in position and magnitude and, as well as generating discrete tone noise, this mechanism produces broad band noise (21).

When the flow is unsteady with distortions in time and space then the noise generated by the interaction with fan blades is random and broad band, though, if the turbulence is sufficiently coherent such that a single eddy is intercepted by a number of blades then discrete components are also generated (21).

Two strategies can be adopted to minimise the influence of turbulence. One is to design the fan so as to minimise its response to turbulence. The other is to minimise the turbulence incident onto the fan. The fan designer can choose elements of both strategies; the system designer must attempt to minimise turbulent or asymmetric inflow.

A number of proposals have been made for designing fans to have minimum response to turbulent or non-axisymmetric flows. The response of a blade to a turbulent transverse gust is typified by the Sears function. However, the response is modified if the blade is cambered and if there is a chordwise gust velocity component (18). Under certain conditions the responses to chordwise and transverse gusts can be made to partially cancel one another. Gearhardt et al (22), Horlock (18) and others have examined some of the design implications but it appears that this technique offers only limited success. Hay (23) has proposed a somewhat different technique, selecting a blade section which, at the design duty, will operate well away from stall.

The literature abounds with reports (24) of successful designs of fan in which alterations to the blade section, chord, camber, blade loading etc have resulted in significant noise reductions in particular circumstances. Seldom, however, does a modification which has been successful on one fan result in similar success on a second fan.

Many researchers have reported reductions in noise through the elimination of sources of turbulence. Few axial or mixed flow fans are now designed with inlet guide vanes. Because the relative flow velocities onto the downstream rotor stage is inevitably high, the wakes from upstream vanes incident onto the rotor blades generate high tonal sound power levels. In the design of axial fans with outlet vanes, the axial separation of the blade rows and the choice of blade numbers can be selected to minimise noise generation. If some form of support is required upstream of the rotor stage it is preferable that the supports are not arranged radially but configured in a chordwise or skewed pattern. This has the effect that a blade does not encounter a disturbance more or less in-phase along its complete span.

Some of the most spectacular reductions have been achieved in the aero engine industry. Special, hemispherically-shaped honeycomb and gauze screens have been designed to fit around the inlets of the large, high by-pass, fan jet engine, ground test beds to minimise the natural turbulence present in the atmosphere being drawn into the fan blade row (25). Moore (26) reported tonal noise being reduced by 8 dB and broad band noise reductions of 5 dB by bleeding off the duct wall boundary layers upstream of a rotor.

If removal of wall boundary layers is beneficial, can it be assumed that moving the fan blades away from the boundary layer, ie increasing the tip clearance would also help? The answer is no. Virtually all research has shown that noise is minimised by having low tip clearance. The flows at blade tips are complex and it would appear that the noise caused by the interaction of the blade tip with the wall boundary layer is less than that caused by the interaction with the tip vortex.

In automotive radiator cooling applications the impeller is generally mounted on the engine block whilst the fan cowl is supported from the radiator which is fixed to the chassis. To allow for the relative movement of the fan in the cowl, large radial clearances are required and this has the effect of reducing aerodynamic performance and incresing noise generation. Increases of the order of 10 dB (27), may occur with the large clearances required on vehicle fans as compared with the noise produced with tight clearances.

Various measures including the use of rotating tip shrouds (27) differing blade shapes and numbers (28) have been used to minimise this effect. A recent paper by Aoki (29) describes the use of a ring placed upstream of the fan to control the recirculating flows passing through the tip gap. A noise reduction of 5 dBA was achieved.

## 3.2    Fan Noise Transmission

The mathematical description of the radiation of fan noise from an open inlet or outlet (30, 31) or its transmission along a duct (32, 33, 34) is complex. Considerable effort has been expended by aero engine manufacturers in the modelling of sound transmission, but it is fair to say that there has been no commensurate effort in studying low-speed, general-purpose, industrial and ventilating fans. After pioneering work by Cremer (35), Baade (36) and supporting experimental evidence from a number of sources, (37, 38) it is now widely accepted that fan inlet and outlet sound are different, and noise radiated from a free inlet or outlet is different from the noise radiated into a ducted system. Qualitative agreement has been shown between differences of fan noise measured in different duct configurations and theories based on treating the fan as analogous to an electrical generator and the ductwork as analagous to a complex electrical

load. Considerably more research effort is required in this area to develop a full understanding of acoustic transmission processes.

## 4    SILENCING

### 4.1    Determination of standardised silencer attenuation performance

It has long been recognised that the acoustic performance of silencers is dependant on the installation conditions, air flow velocity through the silencer and the modal content of the incident noise. Two different terms 'transmission loss' and 'insertion loss' have been used to differentiate the attenuations which would be measured under specific, controlled, laboratory conditions and those which would be achieved in the installation. Both the draft BSI and ISO standards (39, 40) are very specific in detailing the noise source, the test ducting, and the qualification of the installation in which the silencer is tested.

### 4.2    Different types of silencer

Of the three basic types of silencer - active, reactive and resistive - the resistive is the most commonly used in fan applications. Active silencers rely on loudspeakers generating noise which is out-of-phase with the noise to be silenced. Recent developments in electronics have reduced the cost of active silencers and it is likely that there will be considerable growth in their use over the next few years. One significant advantage of active silencers is that they do not cause a restriction in ductwork and hence there is no pressure loss penalty associated with their installation in an air moving system.

Reactive silencers are generally characterised by a change in volume or cross-section area associated with their incorporation into the duct. Effective attenuators generally require large volumes and they are not at all common in fan systems, though Neise (41) has successfully used resonators fitted to the volute tongues of centrifugal fans.

Absorptive silencers, comprising linings of fibrous or porous foam materials, are the type most typically found in air moving applications. Depending on the application, the lining may be held behind a perforated metal sheet and many have a thin, protective plastic film to prevent contamination of the lining. The acoustic performance of the silencer is determined by a number of factors including the absorption characteristics (or impedance) of the lining material and the cross-section of the lined section. In order to achieve good attenuation the lining thicknesses have to be reasonably thick and the spacing between lined walls reasonably thin. This requirement means that if a lining is applied to a duct wall the free area for air flow is reduced and pressure losses are incurred.

Generally, therefore, silencers are of greater external cross-section than the ducts in which they fit and they often contain additional absorbent material in the form of splitters or pods.

The addition of a silencer into an aero-dynamic circuit has usually one direct influence - that of increasing the system resistance. Particular combinations of silencer design and fan installation configuration are poor. For example, it is not good practice to have a splitter silencer with horizontal baffles at the discharge of a centrifugal fan. The distribution of flow is such that most of the flow will be directed along the top of the silencer causing a large pressure drop and the possibility of high levels of regenerated noise. A better arrangement is to have the splitters arranged vertically or to have a sufficient space between the fan and silencer to allow flow to be more evenly distributed.

## 4.3    Acoustic performance installation effects

The acoustic performance of a silencer is affected by a large number of factors including the modal content of the incident noise, the air flow velocity through the silencer and the impedance of the circuit in which the silencer is placed. Fry (42) has reported a series of investigations on installation effects which affect silencer performance and some examples are quoted below.

### 4.3.1 Variation with silencer length

An example is shown in Fig. 6. A silencer 1.5D long gave an increase in attenuation more or less in line with the increase in length, but there was no significant improvement noticed in extending the length to 2.0D. Fig. 7 shows the variation of sound pressure in a long silencer tested at NEL under no flow conditions. Initially, there was a rapid attenuation as the higher order modes were preferentially absorbed and then the attenuation rate reduced and eventually levelled off. It is important, therefore, to recognise that increasing the length of a silencer will not result in a proportionate increase of attenuation.

### 4.3.2 Connection to fan

The attenuation offered by a silencer was generally improved if it was close-coupled to a fan. This is most probably the result of absorption of the sound energy in the higher order modes which, in the absence of the attenuator, would be transferred to lower-order, propagating modes within a few diameters length of the duct, though the change in impedance loading on the fan will modify its generation and radiation characteristics. In tests at NEL, it has been noted that noise reductions at the outlet of an axial fan, attributable to the fitting of a close coupled discharge silencer, have been accompanied by increased noise levels at the fan inlet side of 10 dB at blade passing frequency.

Fry found that with silencers fitted to the discharge of axial fans there was a consistent trend in which decreases in attenuation correlated with the pitch angle of axial fan blades. An example of this is shown in Fig. 8 It was hypothesised that the effect was related to swirling flow and regeneration but the effect was also present on a fan fitted with outlet guide vanes.

There is still much research required to provide a full understanding of the influences of silencers on fan noise generation and attenuation.

### 4.3.3 Location of silencer within system

Whilst not strictly an 'installation effect' the location of a silencer in a system is very important in the acoustic performance of the system. Many reference books (41, 42) show examples of good and bad practice. For example, if breakout noise was not important in a plant room it would be better to site a silencer close to the delivery point of the air so that any noise generated by flow through the system could be attenuated along with the fan noise in a single silencer. If noise in the plant room was important then it would be necessary to connect the silencer to the fan and also, possibly, have a second silencer close to the delivery point.

## 5    VIBRATION INSTALLATION EFFECTS

### 5.1    Determination of standardised vibration performance

Part 5 of BS 848 has just been published and the test measurements determined in accordance with the specified procedures will form the basis for determining the magnitude of any installation effects.

### 5.2    Vibration installation effects

Vibration in fans is generated by fluctuating forces with the principal sources being aerodynamic forces which cause excitation of the blades, motor housing, casing and attached ductwork and mechanical forces resulting from out-of-balance, rubbing, rolling or other contacts. Changes in vibration can occur in an operating fan for a wide variety of reasons including changes in the nature of the aerodynamic forces, which could be caused by increased turbulence, asymmetric flow or flow separation. Many changes in vibration are associated with a fault such as imbalance of the rotating components, shaft distortion, shaft misalignment or damaged bearings. A very useful guide to using vibration as a fault diagnosis is Reference (45).

Vibration problems can be divided into two main categories - those which represent a threat to the integrity or operation of the fan and those which represent a problem to the operating environment. In the first case the source of the problem needs to be identified and rectified. In the second case the level of vibration on the fan casing or bearing housing may be acceptable in that the integrity of the fan is not in question but the vibration which would be transmitted into the fan support structure or foundations cannot be tolerated by the user. In such cases it may not be possible to achieve low vibration levels without the use of anti-vibration mounts and inertia bases.

Care needs to be exercised in selecting anti-vibration mounts in the design of systems which give adequate vibration isolation, provide a stable platform for the fan and cope with reaction thrusts and torques. Special attention is required for cases where the structure on which the fan is mounted is neither very massive nor rigid. An inappropriate selection of mount can lead to excitation of excessive vibration around the resonant frequency of the foundations. A clear exposition of this effect is given by Sharland (44) from which Fig. 9 is taken. An interesting example of a vibration problem caused by the resilience of the floor on which the fan was installed is described by Fermer and Lack (46).

In some applications, such as in semiconductor plants, where very low levels of vibration are being specified, it is found that the levels cannot be met without in situ balancing, the actual process of installing the fan being sufficient to cause imballance in previously factory-balanced units.

It is important that a fan is not operated such that the blade passing frequency or one of its dominant harmonics is close to a natural vibration frequency of a blade. This is usually of importance where a fan can be operated over a range of speeds so that the running frequency may coincide with a natural frequency. Under these conditions large amplitude vibrations and stresses can be built up leading to fracture of the blade. Blade natural frequencies can be quite easily measured with the fan stationary by attaching a lightweight accelerometer to the blade and gently tapping the blade. However, the natural frequency may change with fan operation as the centrifugal and aerodynamic loads can add considerable stiffness, particularly to blades with a long span. Change of operating conditions, particularly the thermal environment, can result in alterations of the natural frequency.

6    INSTABILITIES IN FANS AND FAN SYSTEMS

6.1    Influence of shape of fan characteristic

The operating point of a fan is determined by the intersection of the fan characteristic curve with the system resistance line. Any uncertainty in the determination of the system resistance can be represented by upper and lower limits as indicated in Fig. 10. The effect of the uncertainty on the operating flowrate depends on the relative slopes of the fan characteristic and resistance lines. If the fan characteristic is steep, then the possible variation in flowrate is minimised. Conversely, if the characteristic is flat, then the effect of uncertainty in system resistance characteristic on operating flowrate is potentially large. On critical installations where the resistance is uncertain, selection of a fan with a steep characteristic is obviously advised.

The shape of a fan characteristic is determined by its aerodynamic design. Fans can be designed to have negatively sloping character-istics over their entire flow range (Wilson, (44), Farrant (48)) and, at low pitch angles, most axial fans display negatively sloping characteristics. The exact mechanisms which produce a continuously rising characteristic are not yet fully understood ((49) Carey et al) but they are related to changes in the way in which the pressure is developed by the impeller and to secondary flow patterns established within the fan, as well as to characteristics of any guide

vane stage and the volute or casing.

As flowrate is reduced in axial and mixed flow impellers, the dominant generating mechanism changes from being predominantly blade shape work (ie deflection or diffusion) to centrifugal work (ie radial displacement of the streamlines through the impeller). In axial impellers, in particular, the onset of a positive slope can generally be related to the onset of aerodynamic stalling of the blade or section of the blade. If the blade stall occurs along the whole span then the characteristic will tend towards that shown in Fig. 11a with a very marked decrease in pressure generation.

If the stall is graded, with the tip sections stalling before the hub section, then the characteristic may tend towards that shown in Fig.11c. In this case, if the fan is operated in the stalled condition, the through flow will be stable and, provided the aerodynamic forces on the blades are not excessive, the fan will operate stably. If the characteristic is as shown in Fig. 11a, then it is possible for the fan to operate stably at the points A and B. NEL has had experience of an installation of this type. If some momentary event so increased the system resistance that the flowrate was reduced below that corresponding to the peak pressure point the fan would move to the operating point A. It was possible, by opening a flap in the ductwork to reduce the system resistance for a short period of time to that shown on the line OC. Closing the flap then allowed the fan to move along its characteristic to the stable point B.

If the characteristic is like that shown in Fig. 11b, then stable operation is generally not possible at flows less than that at point X. Surging will occur with the flow through the fan reversing periodically as the fan duty point oscillates between point X and Y. The frequency at which the flow oscillation occurs is affected by the volume and layout of the system in which the fan is operating but is typically of the order of 1 to 10 Hz. When surging occurs the system cannot be operated.

6.2    Fans and stalling

6.2.1 Axial flow fans

It is not recommended practice to operate fans, particularly axial fans, in stall. The fluctuating forces on the blades are high and can induce fatigue resulting in catastrophic failure. Depending on the design of the fan, failure may occur after only a few minutes of stalled operation or the fan may be able to run indefinitely. Just prior to all the blades stalling, the fan may move into a regime where a rotating stall cell or cells exist. In this condition, instead of all the blades stalling simultaneously, a small number of blades stall. The flow distortion introduced by this is such as to cause the stall cell to move around the rotor blades. This phenomenon has been widely studied (50, 51) in axial compressors of aero engines as it is often a precursor to the onset of surge.

A number of fan casing treatments (50, 52) have been developed which delay the onset of stall in axial fans. These treatments appear to establish a stable recirculating flow pattern a little upstream and within the impeller blade tip region.

### 6.2.3 Centrifugal fans

The flow over the suction surface of a centrifugal impeller blade is generally separated and aerodynamic stalling of the blade is not generally a problem. However, rotating stall can be a significant problem (53, 54). The symptom of rotating stall is generally a fluctuation of the fan pressure at a frequency of approximately 2/3 or 4/3 of the shaft rotational frequency and is usually accompanied by an increase in noise and vibration. The magnitude of the pressure fluctuation can range from being imperceptible to being sufficiently great as to damage the fan casing and ductwork and render normal operation of the system impossible. The flowrate through the fan is sensibly constant during operation with rotating stall.

The onset of rotating stall is not simply related to excessively high incidence of the flow onto the blades. Fans tested at NEL have developed severe rotating stall at flowrates as high as 85 per cent of best efficiency flowrate and we have found evidence of rotating stall cells at flowrates above 120 per cent of best efficiency flowrate. Recent research (55) by Professor Railly at Birmingham University has provided insight into rotating stall and he has designed an impeller which was free of rotating stall. Further work is, however, still necessary to develop a method of predicting whether a particular design of fan will be prone to operate in a rotating stall regime and produce unacceptably high pressure fluctuations.

It has been found that rotating stall is affected by the system in which the fan operates and the flow conditions into and out of the impeller. Wright et al (56) report changes in the onset of stall with severely distorted inlet air flow. Railly (55) found that when an impeller was throttled by a set of variable vanes located circumferentially around the impeller exit, rotating stall could be prevented although this cannot be implemented as a practical solution.

### 6.3 Inlet vortex instability in centrifugal fans

Many centrifugal fans are fitted with variable pitch inlet guide vanes to provide a means of altering the performance characteristic. These vanes are generally either fitted close to the inlet eye of the impeller or at the entry to the inlet box. When the vanes close they generate a swirling flow. If the swirl rotation is in the same sense as the impeller rotation, closing the vanes reduces the pressure rise and power consumption at a given flowrate. It has been found that as the vanes close to generate increased swirl, severe pressure pulsations and vibration can occur. The effects can be sufficient to prevent operation of the machine. The instability is associated with the core of the swirl vortex becoming unstable (57, 58) and precessing around the fan impeller. The dominant frequency of the pulsations is of the order of one to two times shaft rate.

Various devices have been used to reduce the amplitude of the disturbances. The most commonly used are dorsal fins which are small radial fins mounted almost within the impeller eye and downstream of the variable guide vanes. These fins remove some of the swirl from the core of the vortex and the vortex then tends to stabilise round a central axial jet flow. Experimentally, it has also been found that a pronounced asymmetry in the inlet can prevent the precession of the vortex. In the case where a fan has an inlet box, a splitter plate in the box can stabilise the vortex.

### 6.3 Fan operation in parallel

When two or more fans are operated in parallel it is possible to develop unstable flow conditions. These may take a number of different forms. One such is 'hunting' where there is a regular variation in the flow through both fans with first one fan passing more flow than the second and then the position being reversed. The dynamics of the system and the pressure versus flowrate characteristics of the fans influence the magnitude and frequency of the flowrate (and pressure) excursions.

Where fans operating in parallel have an aerodynamic performance characteristic with a positive slope, so that, for a given static pressure difference the fan can operate at two flowrates, problems can arise, particularly during start-up or following sudden changes of load. It is then possible that one of the fans will operate at a flowrate above the stall line while the other fan will operate in a stalled condition. A detailed coverage of this type of operation is given in Reference (13).

### 7 SAFETY ASPECTS

It is the duty of a manufacturer of equipment to ensure that it is safe under normal operating conditions. Fans already have to comply with various mandatory and voluntary codes covering electrical safety, guarding, noise etc. Under forthcoming EEC legislation, see Reference (59), it will be a requirement for a manufacturer to consider and be responsible for a much wider range of aspects. The scope of the regulations states that "when designing machinery and drafting instructions, the manufacturer must envisage not only the normal use of the machinery but also uses which could be reasonably expected. The machinery must be designed to prevent abnormal use if such use would engender a risk. In other cases the instructions must draw the user's attention to ways - which experience has shown might occur - in which the machinery should not be used".

The implications of these regulations for fan manufacturers and system designers are very significant. Manufacturers will seldom have knowledge of the systems into which their fans are to be installed. As this review has shown, there are interactions between the fan and the system which can result in the fan being operated well away from its designed or specified duty and that, in certain circumstances, this can be damaging to the integrity of the fan. Manufacturers, system designers, installers and operators are going to have to co-operate much more closely than before to ensure safe operation of fans and fan-related plant.

## 8 CONCLUSIONS

This paper has reviewed some of the changes in fan performance caused by the interaction of the fan with the system in which it operates.

The importance of installation effects has been highlighted in terms of commercial pressures on system designers and component manufacturers, unnecessary energy loss which affects users and compliance with safety legislation. It is clear that a lot of further research work is required. While much of the work will, of necessity, be experimental, developments in Computational Fluid Dynamics should provide opportunity for detailed study of the secondary flows which account for many of the installation effects.

The information which the research will produce will ultimately be of use and benefit to manufacturers, consultants, designers and users. It is only sensible that there is co-operation between the interested parties so that there is no unnecessary duplication of effort. This is beginning to happen in, for example, the collaborative projects organised by the FMA on fan installation effects and the EEC on pressure losses in ductwork components. The pooling of resources at both National and International levels will result in the expeditious generation of the necessary information.

ACKNOWLEDGEMENTS

This paper is contributed with the permission of the Director, National Engineering Laboratory, Department of Trade and Industry. It is Crown copyright.

## REFERENCES

(1)  GUTTLER, G.  European catalogue of energy loss coefficients of components used in air distribution systems, cost benefit analysis.  SWD Software Design GmbH, Frankfurt am Main. 1985.

(2)  BRITISH STANDARDS INSTITUTION.  Fans for general purposes.
Part 1:  Methods of testing performance
Part 2:  Methods of noise testing
Part 5:  Methods of vibration measurement.

(3)  INTERNATIONAL STANDARDS ORGANISATION.  ISO DP 5801.  Method of fan performance testing.

(4)  ANON.  Fans and systems.  AMC Publication 201.  Air Moving and Control Association Inc., 30 West University Drive, Arlington Heights, Illinois, USA.

(5)  DEEPROSE, W. M., SMITH, T. W.  The usefulness of BS 848 : Part 1 : 1980 in estabishing the installed performance of a fan.  Paper C110/84.  I.Mech.E. Conf. on Installation Effects in Ducted Fan Systems, May, London 1984.

(6)  CORY, W. T. W.  The effects of inlet conditions on the performance of a high specific speed centrifugal fan.  Paper C115/84.  I.Mech.E. Conf. on Installation Effects in Ducted Fan Systems, May, London 1984.

(7)  ZALESKI, R. H.  System Effect Factors for axial flow fans.  AMCA Engineering Paper 20011-88-A2, March, 1988.

(8)  HESSELGREAVES, J. E.  A correlation of tip clearance/efficiency measurements on mixed and axial flow turbomachines.  Report 423.  East Kilbride, Glasgow: National Engineering Laboratory, 1970.

(9)  NIXON, R. A. and CAIRNEY, W. D.  Scale effects in centrifugal cooling water pumps for thermal power stations.  NEL Report 505.  East Kilbride, Glasgow: National Engineering Laboratory, 1972.

(10)  WRIGHT, T.  Centrifugal fan performance with inlet clearance.  J. Engng for Gas Turbine and Power, Trans ASME 106, pp 906-912 October, 1984.

(11)  BECIRSPAHIC, S.  Fan system effect - work carried out in CETIAT.  AMCA Engineering Paper 20010-88-AO, March, 1988.

(12)  MILLER, D. S.  Internal flow systems.  BHRA Fluid Engineering Centre, Cranfield, 1986.

(13)  JORGENSEN, R.  (ed) Fan Engineering, Eighth Edition.  Buffalo Forge Company, Buffalo, New York, 1983.

(14)  ANON.  Air Systems.  AMCA Publication 200.  Air moving and Control Association Inc., 30 West University Drive, Arlington Heights, Illinois, USA

(15)  FFOWCES-WILLIAMS, J. E. and HAWKINS, D. L.  Theory relating to the noise of rotating machinery.  J. Sound and Vibr. 1969, Vol 10, No 1, pp 10-21.

(16)  MARGETTS, E. J.  A demonstration that an axial fan in a ducted inlet ducted outlet configuration generates predominantly dipole noise.  J. Sound and Vibr., 1987, 117(2), pp 399-406.

(17)  HORLOCK, J. H.  Fluctuating lift forces on aerofoils moving through transverse and chordwise gusts.  Trans ASME. J. Basic Engng, December 1986, pp 494-500.

(18)  NAUMANN, H, YEH, H.  Lift and pressure fluctuations of a cambered aerofoil under periodic gusts and applications in turbomachinery.  Trans ASME, J. Engng Pwr. January 1973, pp 1-10.

(19)  SHARLAND, I. J.  Sources of noise in axial flow fans.  J. Sound and Vibr., 1964, 1(3), pp 302-332.

(20)  WHITFIELD, C. E. and HAWKINS, D. L.  An investigation of rotor noise generation by aerodynamic disturbance.  Paper C105/75.  Conf. on Vibrations and Noise in Pump, Fan and Compressor Installations.  I.Mech.E, London, September 1975.

(21) HANSON, D. B. Spectrum of rotor noise caused by atmospheric turbulence. J. acoust. Soc. Am., July 1974, Vol. 56, No 1.

(22) GEARHARDT, W. S., HENDERSON, R. E., MCMAHON, J. F., HORLOCK, J. H. The quasi-steady design of a compressor or pump stage for minimum fluctuating lift. Trans ASME, J. Engng Pwr. Paper 68 – WA/FE-12.

(23) HAY, N., MATHER, J. S. B., METCALFE, R. Fan blade selection for low noise. I.Mech.E Seminar on Industrial Fans, April 1987, London.

(24) NEISE, W. Fan noise generation mechanisms and control methods. Proceedings Internoise 88, Avignon, France pp 767–775.

(25) PERACCHIO, A. A. Assessment of inflow control structure effectiveness and design system development. J. Aircraft, December 1982, Vol. 19, No 12.

(26) MOORE, C. J. Reduction of fan noise by annulus boundary layer removal. Paper 73, ANC 4. British Acoustical Society, Spring Meeting, April 1973.

(27) LAISE, T. D., MELLIN, R. C., LONGHOUSE, R. E., and PRYJMAK, B. I. Electric motor cooling fan with high ram airflow – a fuel economy improvement. SAE 790722 Society of Automotive Engineers, 1979.

(28) HOFE, R. V. and THIEN, G. E. An efficient approach to design low noise automotive cooling systems. Paper C 134/84. I.Mech.E Conference on Vehicle Noise and Vibration, London, June 1984.

(29) AOKI, Y, MATSUDA, K., TOMINAGA, T., KONDO, F., YAMAGUCHI, N., and HAYASHI, M. Noise reduction of condenser cooling fans for automotive air conditioners. Mitsubishi Heavy Industries Ltd, Technical Review, Vol. 24, No 3, October 1987.

(30) TYLER, J. M. and SOFFRIN, T. G. Axial flow compressor noise studies. SAE Trans., 1962, Vol. 70 pp 309–332. New York: Society of Automative Engineers.

(31) LOWSON, M. V. Theoretical studies of compressor noise NASA CR 1287. March 1969.

(32) MORSE, P. and INGARD, K. U. Theoretical acoustics. McGraw-Hill, London, 1968.

(33) MECHEL, F. P. The computation of circular silencers (in German). Acustica, 1976, Vol. 35 pp 179–189.

(34) DAVIES, P. O. A. L. Practical flow duct acoustics. J. Sound and Vibr., 1988, 124(1) pp 91–115.

(35) CREMER, L. The treatment of fans as black boxes. Second Annual Fairey Lecture. J. Sound & Vibr., 1971, 16(1), 1–15.

(36) BAADE, P. Effects of acoustic loading on axial flow fan noise generation. Noise Control Engineering, January–February 1977, 8(1), 5–15.

(37) WOLLHERR, H. Acoustic investigations of radial fans using 4-pole theory (in German). PhD Thesis, Technical University of Berlin, 1973.

(38) BOLTON, A. N. and MARGETTS, E. J. The influence of impedance on fan sound power. Paper C124/84. I.Mech.E. Conf. on Installation Effects in Ducted Fan Systems, May, London 1984.

(39) BRITISH STANDARD INSTITUTION. BS 4718. Methods of test for silencers for Air Distribution Systems, 1971 (under revision).

(40) INTERNATIONAL STANDARDS ORGANISATION. ISO 7235 acoustics – measurement procedures for ducted silencers (publication anticipated 1990).

(41) NEISE, W. and KOOPMANN, G. H. Reduction of centrifugal fan noise by use of resonators. J Sound & Vibr., 1980, Vol 73, pp 297–308.

(42) FRY, A. T. On fan application of noise attenuators. Paper C123/84. I.Mech.E. Conf. on Installation Effects in Ducted Fan Systems, May, London 1984.

(43) WEBB, J. D. (ed). Noise Control in Mechanical Services. Sound Attenuators Ltd and Sound Research Laboratories Ltd, Colchester, Essex, 1972.

(44) SHARLAND, I. Woods practical guide to noise control. Woods of Colchester Ltd, Colchester, 1986.

(45) Michael Neale and Associates. A guide to condition monitoring of machinery. HMSO, London, 1979.

(46) Fermer K-E, and Lack C. W. Vibration isolation of fans considering the elastic properties of the floor. Paper C116/84. Conf. on Installation Effects in Ducted Fan Systems, I.Mech.E, London, May 1984.

(48) FARRANT, P. E. An approach to mixed flow fan design. Fluid Mechanics Silver Jubilee Conference. Paper 2.1. East Kilbride, Glasgow: National Engineering Laboratory, 1976.

(50) GREITZER, E. M. Review – Axial compressor stall phenomena. Trans ASME, J. Fluids Engineering, June 1980, Vol 102, pp 134–151.

(51) TAKATA, H. and NAGANO, S. Non-linear analysis of rotating stall. J. Engng Pwr. Trans ASME, Series A, Vol. 94, No 4, October 1972.

(52) BARD. H. The stabalisation of axial fan performance. Flakt Industry, AB. Vaxjo, Sweden.

(53) WORMLEY, D. N., ROWELL, D. and GOLDSCHMEID, F. R. Air/gas system dynamics of fossil fuel power plants - pulsations. EPRI Report CS 2206.

(54) BOLTON, A. N. Pressure pulsations and rotating stall in centrifugal fans. I.Mech.E. Conference on Vibrations and Noise in Pump, Fan and Compressor Installations, September 1975.

(55) RAILLY, J. W. and EKEROL, H. Influence of a closely coupled throttle on the stalling behaviour of a radial compressor stage. Trans ASME, J. Engng for Gas Turbines and Power, April 1985, Vol. 107, pp 513-527.

(56) WRIGHT, T., MADHAVAN, S., DiRe, J. Centrifugal fan performance with distorted inflows. Trans ASME, J. Engng for Gas Turbines and Power, October 1984, Vol. 106, pp 895-900.

(57) PEARSALL, I. S., and BOLTON, A. N. Fluid factors causing failure. Paper C159/80. I.Mech.E. Convention on Fluid Machinery Failures - Prediction, analysis and prevention. April 1980.

(58) KIMURA, Y., HIGASHIMORI, H., WATABIKI, N. and INADA, N. Research on the similarities and the solutions of an inlet cone vortex in centrifugal fans. Trans Japan Soc. Mech. Engng, 1987, Vol. 53, Part 488, pp 1262-1269, 1987.

(59) COUNCIL OF EUROPEAN COMMUNITIES. Common position adopted by the council on ...... with a view to the adoption of a council directive on the approximation of the laws of the member states relating to machinery. Document 10231/88, Brussels, December 1988.

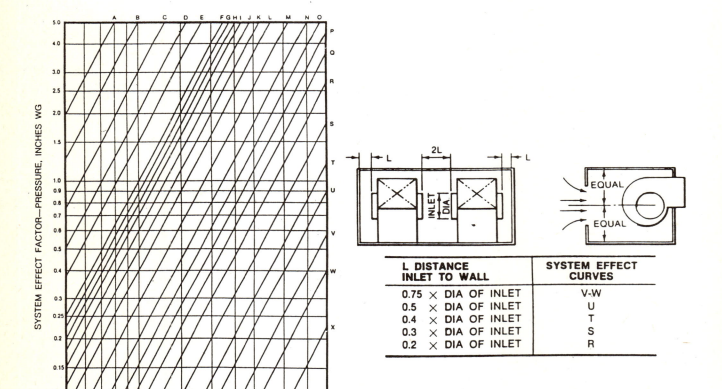

Fig 2b  Example showing use of system effect curves

| L DISTANCE INLET TO WALL | SYSTEM EFFECT CURVES |
|---|---|
| 0.75 × DIA OF INLET | V-W |
| 0.5 × DIA OF INLET | U |
| 0.4 × DIA OF INLET | T |
| 0.3 × DIA OF INLET | S |
| 0.2 × DIA OF INLET | R |

Air Density = 0.075 lb per cu ft

Fig 2a  AMCA system effect factor curves

**PARALLEL BLADED DAMPER ILLUSTRATING DIVERTED FLOW**

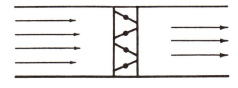
**OPPOSED BLADED DAMPER ILLUSTRATING NON-DIVERTING FLOW**

Parallel vs. Opposed Dampers

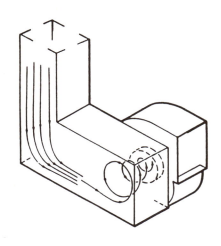

Example of a Forced Inlet Vortex
(Spin–swirl)

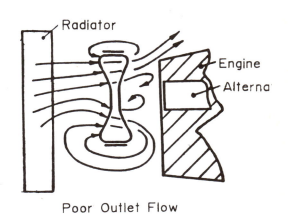

Poor Outlet Flow

Fan in a Vehicle

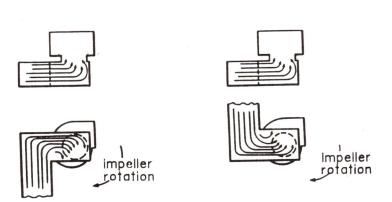

Inlet Duct Connections Causing Inlet Spin

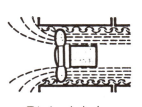

Plain Inlet

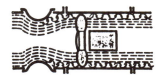

Slack Flexible Connector

Impeller Immediately
After Bend

Fig 1    Examples of installations likely to produce aerodynamic or noise
installation effects

! Pressure tapping
! Resistance thermometer
✦ Traverse location

Direction of flow

All dimensions in mm

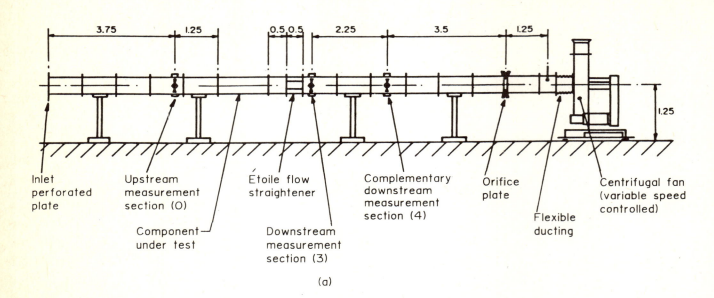

Inlet perforated plate

Upstream measurement section (O)

Component under test

Étoile flow straightener

Downstream measurement section (3)

Complementary downstream measurement section (4)

Orifice plate

Flexible ducting

Centrifugal fan (variable speed controlled)

(a)

Fig 3a     Determination of component pressure loss — test rig

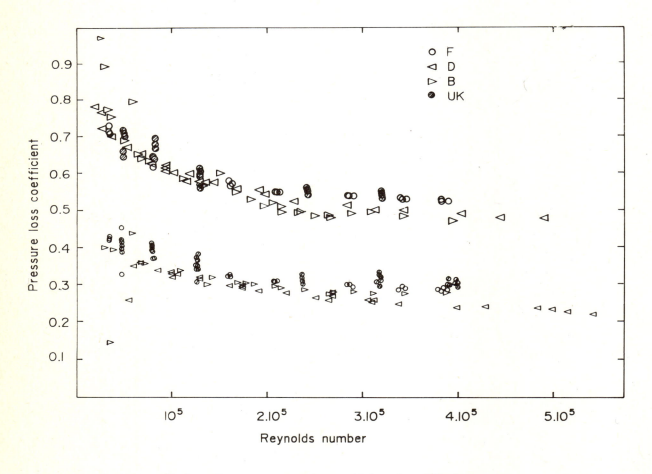

Fig 3b     Measured pressure loss coefficients for a 45° elbow and two 90° elbows in two planes

14

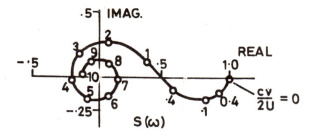

Fig 4     Sears response function

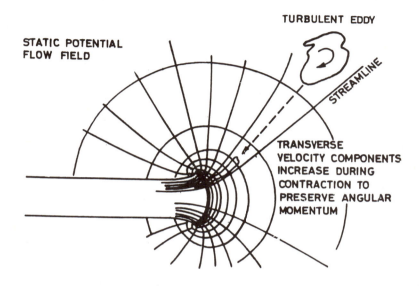

Fig 5     Elongation of atmospheric turbulence at fan inlet

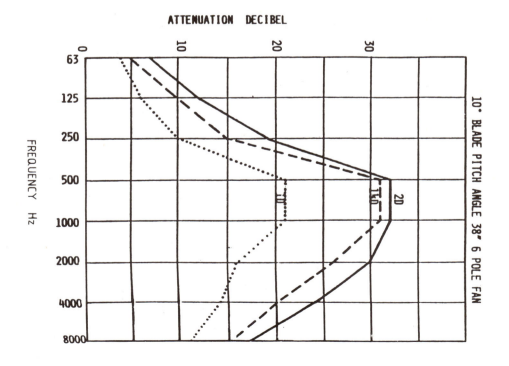

Fig 6     Influence of silencer length on attenuation

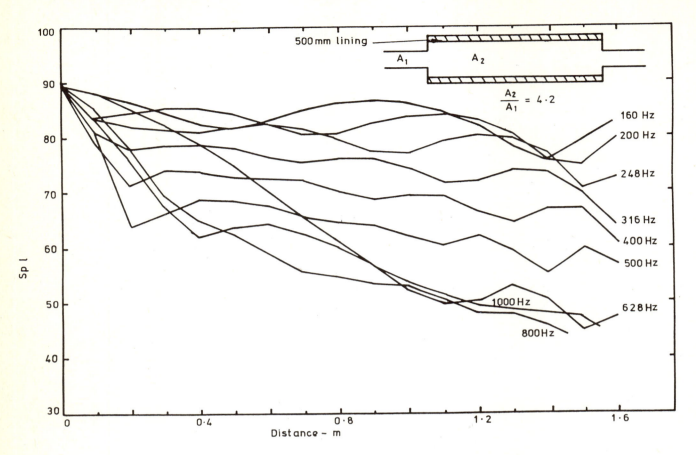

Fig 7     Variation of sound pressure levels along the internal axis of a
          silencer

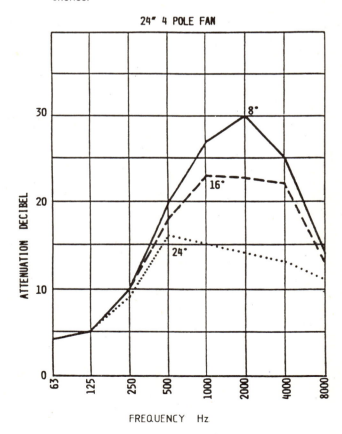

Fig 8     Influence of blade pitch on silencer attenuation

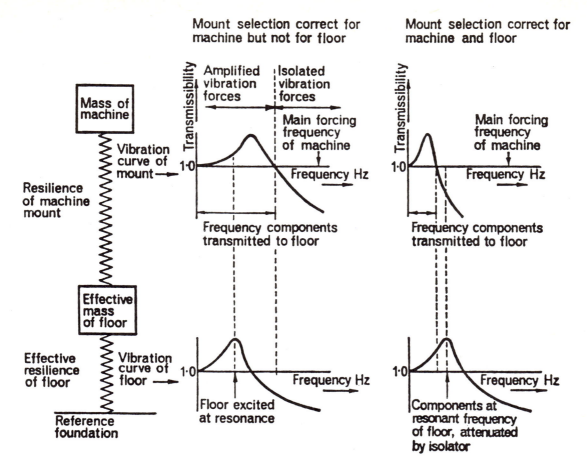

Fig 9      Influence of resilient foundation on vibration isolation

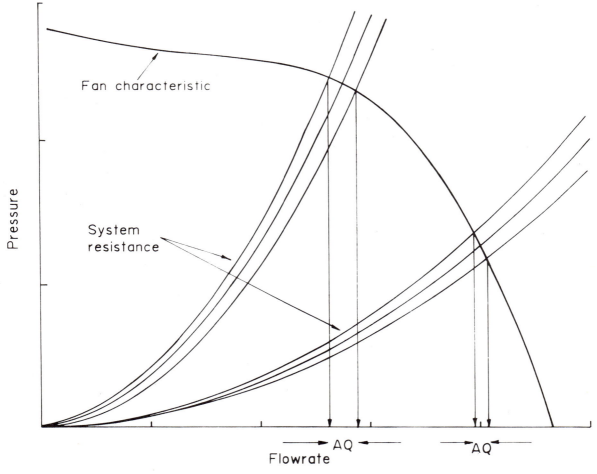

Fig 10      Influence of slope of fan characteristic on operating flowrate
for 10 per cent change of system resistance

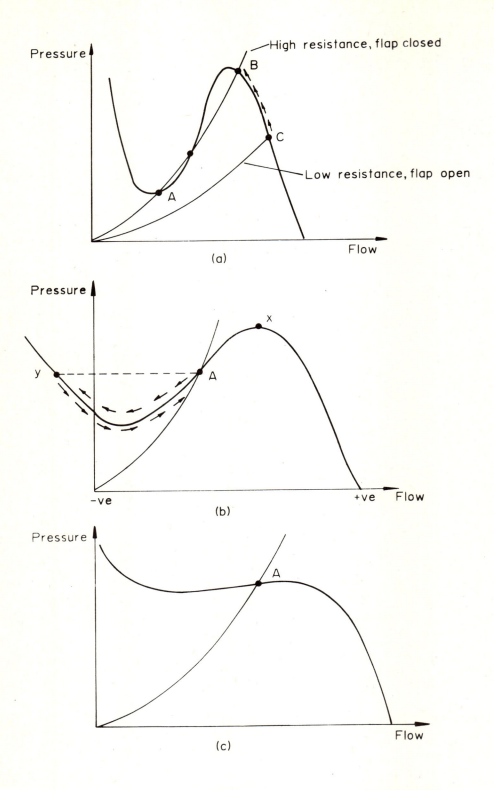

Fig 11    Influence of shape of fan characteristic on aerodynamic stability

# C401/017

# Fans in air handling units

**W R WOODS BALLARD**, MSc, CEng, MIMechE
Woods of Colchester Limited, Colchester

Much has been surmised concerning the so-called 'system effect' of operating fans in air handling unit installations.  Though catalogue data is based on standardised laboratory tests, usually in accordance with BS848, installed performance often falls short cf predictions.  Laboratory tests have been undertaken to gain a better measure of the 'system effect' for different types of fans.

The paper presents the results from tests on centrifugal and mixed flow fans and discusses their suitability for use in air handling units.

## 1    INTRODUCTION

Traditionally, and for good practical reasons, centrifugal fans are most frequently chosen as the air moving device for air handling units. The double inlet type, particularly, is considered to have a relatively low noise level for a given duty, as it can handle an acceptable flow rate at the same time as developing a high pressure at a low tip speed.  Static efficiencies are allegedly high and the performance of belt-drive units can be readily adjusted on site.  A centrifugal fan can also be arranged to discharge horizontally, in the axial direction, or vertically, up or down.  In the vertical case, space and a duct bend, which would be necessary if the air were discharged horizontally before being turned vertically, is saved.

Though they have some installation advantages over the centrifugal, axial flow fans are less often specified for air handling units. When they are, it is usually because a variable blade pitch fan is preferred.  At the lower flows however, below about 5m³/s, the great majority of air handling unit fans are centrifugals.

With advances in turbomachinery design and manufacturing techniques, mixed flow fans have become commercially available.  These are now being manufactured in significant numbers, many for installation in air handling units.  Since the mixed flow fan incorporates a number of the advantages, both from performance and installation points of view, of both centrifugal and axial flow fans, an appropriate programme of tests was undertaken.

The programme looked at the installed performance of a backward curved bladed and a forward curved bladed centrifugal and a mixed flow fan, all selected to provide the same duty in the air handling unit.  Performance was compared with catalogue data, upstream and downstream proximity effects were examined and measurements of downstream silencer performance were taken.

## 2    FANS TESTED

500mm in-line belt drive mixed flow fan running at 2540 r.p.m.

355 backward curved blade, DWDI, belt drive, centrifugal fan running at 2400 r.p.m.

420mm forward curved blade, DWDI belt drive, centrifugal fan running at 1150 r.p.m.

Fan duty; flowrate of 2.1m³/s at a pressure of 700 Pa.

## 3    TEST CONFIGURATIONS

Initially, the fans were tested to BS848:Part 1: 1980.  The test configuration used was that most appropriate to an air handling unit, namely Type A.  The Type A arrangement is intended to simulate an installation where there is no upstream or downstream duct connection, in other words, the fan draws from and exhausts into an area of relatively low mean velocity.

The fans were also tested in a typical air handling unit with a cross-sectional area of a little less than 1 m² and a coil face velocity of 2.3 m/s.  Fan pressure was taken from the pressure differential across the fan plate and the flow rate was measured using a BS848:Part 1: 1980 conical entry.  The fan duty point was altered by changing the downstream resistance all as indicated on Fig 1a.

Sound levels downstream of the fan were established using a calibrated test section constructed in accordance with BS848:Part 2:1985 and shown on Fig 1b.

## 4    COMPARISON OF CATALOGUE AIR PERFORMANCE

Figures 2a, b and c show the catalogue and test performance of the three fans which were the subject of the test programme.  All the base catalogue corrections for the absence of duct connections has been used.

Fig 2a illustrates the mixed flow fan performance.  This was an early prototype but shows close agreement with catalogue data except towards maximum pressure where there was a corresponding reduction in absorbed power.  Apart from the two lowest flow test points, test data was well within BS848:Part 1, Class 'B' tolerance.

Performance of the backward curved bladed centrifugal is given on Fig 2b. Similarly to the mixed flow fan, the two lowest flow test points fall outside the Class 'B' tolerance. The measured absorbed power, except at around 1 m³/s flowrate, bore little resemblance to the catalogue in either level or shape. The reason for this discrepancy was not established, however the same calibrated motor was used to drive all three fans.

Fig 2c shows the same data for a forward curved bladed centrifugal. Good agreement is shown between catalogue and test data.

## 5  COMPARISON OF TEST AND INSTALLED AIR PERFORMANCE

In Figures 3a, b and c, the BS848 Type A test performances are compared with those in accordance with Fig 1a. For all three fans, it can be stated that the AHU test performances show, in general, a lower pressure at the higher flow rates and a higher pressure at the lower flow rates. This consistent pattern is unlikely to be due to a single factor. It is probably a combination of test configuration and modifications to the flow at the fan inlet and discharge resulting from an air handling unit installation.

Though not shown separately, the air handling unit test results, for all three different types of fan, do show agreement, well within BS848 Class 'B' tolerance, with the manufacturers' data for their fans without a discharge duct. This evidence should provide some comfort to air handling unit manufacturers and fan manufacturers alike. A word of caution is necessary however. Though good agreement between catalogue and installed test data has been demonstrated for three very different fans, large differences in discharge geometry or installation clearances, from those tested, could give different results.

As a reference, the mixed flow fan has a hub/duct outlet area ratio of 0.5 and the two centrifugal fans had a throat to outlet flange area ratio of 0.6. The air handling unit used for the tests illustrated on Fig 3 included an upstream simulated coil with a pressure drop of 60 Pa at a flowrate of 2.1 m³/s. Downstream, a 1.25m long splitter silencer was installed 250mm from the fan discharge, but no diffusing screens were fitted.

## 6  UPSTREAM INSTALLATION EFFECTS

As the mixed flow fan was a relative newcomer to the air handling unit market, its sensitivity to the particular installation requirements of an air handling unit was assessed with special reference to the upstream and downstream proximity of typical components.

To determine the effect of the upstream proximity of typical components, a simulated heater coil was used. This consisted of sheets of perforated steel plates selected to give a pressure drop of 60 Pa at a flowrate of 2.1 m³/s. It was positioned 250mm and 125mm upstream of the fan diaphragm plate and the fan performance measured as in Fig 1a. The velocity distribution 50mm upstream of the coil was also measured using a hot wire anemometer. The pressure drop of 60 Pa was chosen so that it would not have an unduly large effect on the velocity distribution as the higher the pressure drop the more uniform the velocity profile.

The test results showed that, contrary to what might have been expected from a fan with an axial inlet parallel with the airflow, there was no evident increase in velocity in line with the fan inlet. What was demonstrated was that the conditions or components upstream of the simulated coil have a far greater effect on the velocity distribution across the coil than the proximity of the downstream fan. High velocities measured at the sides of the coil were directly in line with the upstream silencer airways.

The effect on fan performance of the upstream proximity of the coil was hardly measurable. For this particular installation where the simulated coil pressure drop and the fan dynamic pressure were the same, it was concluded that the mixed flow fan can be positioned 25 per cent of the fan diameter, or 12½ per cent of the coil width/height, from the coil without detriment to the fan performance or the velocity distribution through the coil.

## 7  DOWNSTREAM INSTALLATION EFFECTS

To determine the effects of a downstream silencer on the fan and silencer performance, and to evaluate the usefulness of diffusing screens, a further series of tests were performed.

The velocity distribution was measured at a position corresponding to the inlet to the silencer airways. With a diffusing screen of 30 Pa pressure drop, but without splitters, the velocity profile for the mixed flow fan was fairly uniform. That for the centrifugal fans slightly less so. Further tests showed that the screen does have a beneficial effect on the downstream velocity profile on both fans.

Performance of the mixed flow fan was measured with the screen and silencer fitted 350 and 480mm downstream. Fan performance was also measured after removing the screen and moving the silencer to 270mm downstream of the fan. This had virtually no effect on fan performance.

## 8  SOUND LEVEL COMPARISONS WITH CATALOGUE

Owing to the way in which it is presented, it has been difficult to interpret the acoustic data for the centrifugal fans. However, the information shown on Fig 5 is the same as that which is used by sales estimators for quotation purposes. The mixed flow fan, being of later design, has been subjected to the full rigours of testing to BS848:Part 2:1985 and data specific to the outlet in-duct and free-field sound power spectra is available.

Fig 5, in addition to the catalogue data, shows the results of tests taken in accordance with Fig 1b. In attempting to obtain sound levels for direct comparison with the catalogue data, the measurements were taken without a silencer but with a diffusing screen fitted. Inspite of using a more-or-less laboratory method to measure the fan sound levels as installed in an air handling unit, there is little correlation between catalogue and test data whatever arguments one might use, as the discrepancies are so great - up to 15dB. Acoustically, it could be said that the fan is installed in a sort of semi-reverberant plenum followed by a square-to-round tapered duct. It is quite likely that this arrangement could have a different effect on the noise output of the fan compared with a semi-reverberant room or an anechoic chamber.

Looking at the octave bands which are most significant in respect of air system attenuation namely the 125 and 250Hz octave bands, it will be noted that test figures are higher than catalogue for the two centrifugal fans but the reverse is true for the mixed flow fan. With the exception of the forward curved bladed centrifugal at 250Hz, the mixed flow fan shows the lowest sound levels in the important bottom three octave bands.

Fig 5 also displays the performance of 1.25m long splitter silencer fitted downstream of the diffuser screen. The silencer had 65mm airways and was fitted with half splitter width sideliners. The test data shows that the silencer performed best, in all octave bands, when installed behind the mixed flow fan. Catalogue data for the silencer, established on the basis of tests carried out in accordance with BS4718, is also shown. With the fans, the acoustic and airflow conditions under which the silencer was operating in the air handling unit installation, were quite different to those pertaining in the BS tests. Also it is likely that the original test data was subject to a degree of scaling/extrapolation which would not now be acceptable.

Clearly there is a great deal to be learned concerning the acoustic performance of fans and splitter silencers when installed in air handling units and its relationship to information established by BS848:Part 2 and BS4718.

9    SUMMARY AND CONCLUSIONS

Air performance tests in an air handling unit have shown that the catalogue data for the fans tested gives an accurate indication of the installed performance. Provided data for a Type A installation is used, the System Effect Factor is 1.

Since the three fans tested were substantially different types of fluid machine, the above conclusion should follow for other sizes of machine provided the geometry and velocities are similar.

All three fans tested were suitable for use in air handling units, there being negligable influence on fan performance due to the upstream or downstream proximity of other components.

In the 125 and 250Hz octave bands, the mixed flow fan showed the best outlet sound levels particularly when coupled with a splitter silencer. The silencer performed better with the mixed flow fan than with either of the two centrifugal fans. This may have been due to the more uniform velocity profile measured at the plane of the silencer inlet.

There was little or no correlation between the published sound data for the fans, or the silencer, and that established in the course of this test programme. Though some possible reasons for this have been put forward, there is a clear need to gain a better understanding of the acoustic effects of installing a fan and a silencer in an air handling unit. A dedicated research programme is suggested.

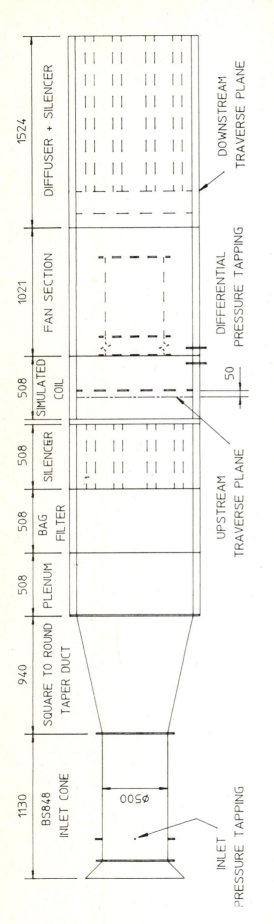

1a PLAN OF AIR HANDLING UNIT TEST DUCT.

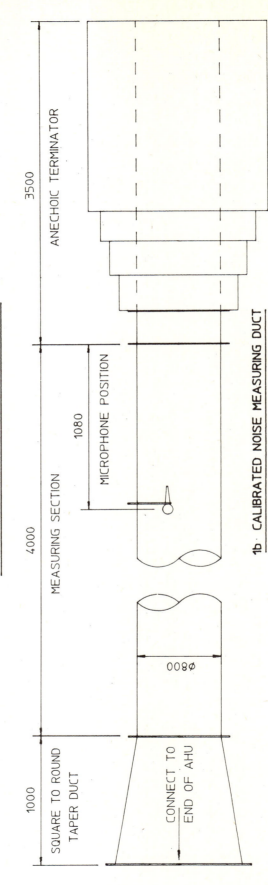

1b CALIBRATED NOISE MEASURING DUCT

Fig 1    Arrangement of test installation

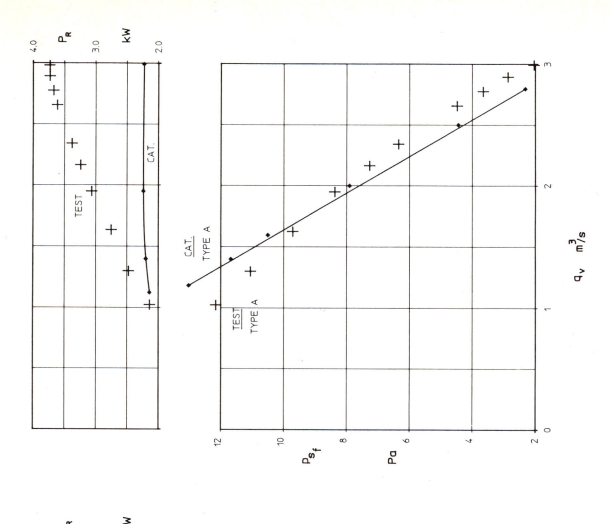

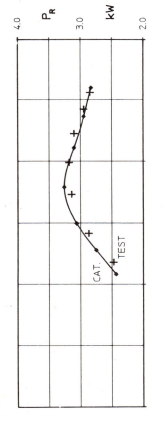

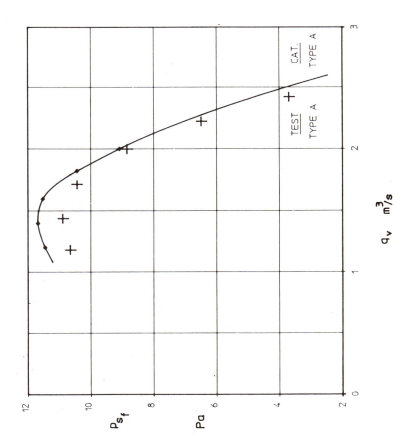

Fig 2b BC centrifugal 2400 r/min

Fig 2a MX50 axcent 2 2540 r/min

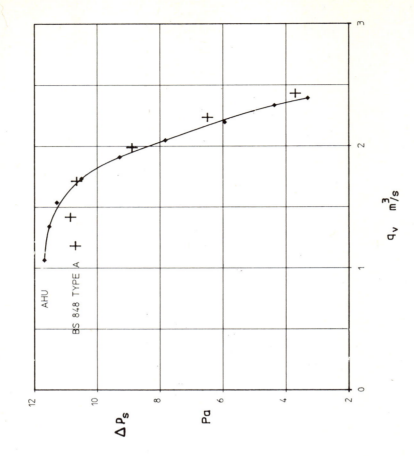

Fig 3a    MX50 axcent 2 2540 r/min

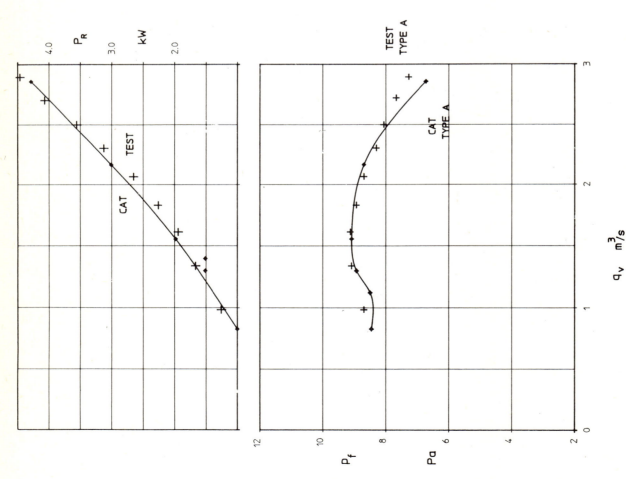

Fig 2c    FC centrifugal 1150 r/min

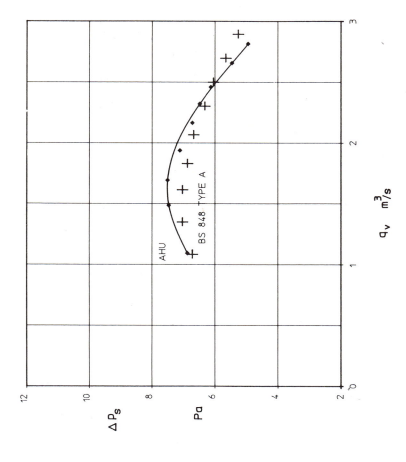

Fig 3c   FC centrifugal 1150 r/min

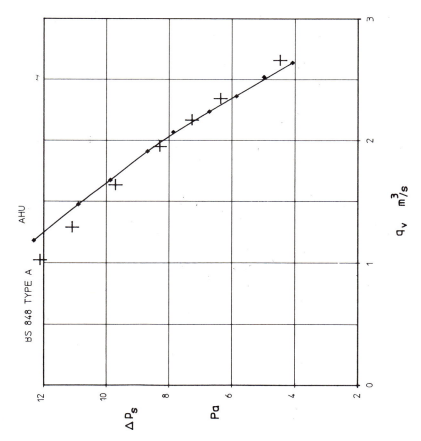

Fig 3b   BC centrifugal 2400 r/min

AIR HANDLING UNIT INSTALLATION
1.25m LONG WITH SIDE LINERS AND
65mm AIRWAYS

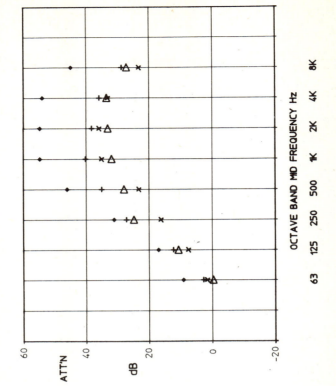

SILENCER CATALOGUE
MIXED FLOW
FC. CENTRIF.
BC. CENTRIF.

OCTAVE BAND MID FREQUENCY Hz

Fig 5    Silencer performance comparisons

FAN SOUND POWER LEVELS
OUTLET 2.1 m³/s
TEST DATA WITH DIFFUSING SCREEN IN.

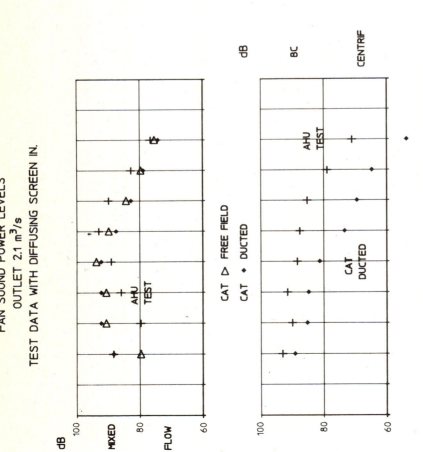

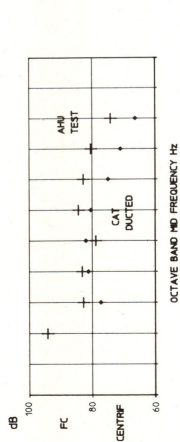

CAT △ FREE FIELD
CAT ◆ DUCTED

OCTAVE BAND MID FREQUENCY Hz

Fig 4    Catalogue comparisons

26

# System effect factors for axial flow fans

**R H ZALESKI**, BSME, ASHRAE, AMCA
Acme Engineering and Manufacturing Corporation, Oklahoma, USA

SYNOPSIS A test program was undertaken by AMCA on axial flow fans to determine system effect factors for varying lengths of outlet duct and the effects of two and four piece mitered elbows installed in close proximity to the test fan inlets and outlets. The duct configurations tested were analyzed for the effects on fan pressure capability, fan power, sound level and fan stability. System effects are provided for both tubeaxial and vaneaxial fans.

## 1. INTRODUCTION

In the summer of 1985, an Air Movement & Control Association (AMCA) technical committee was established to review the AMCA Fan Application Manual, namely AMCA Publications:

201 – <u>Fans and Systems</u>
202 – <u>Troubleshooting</u>
203 – <u>Field Performance Measurements</u>

In the review of AMCA Publication 201 – <u>Fans and Systems</u>, the technical committee came to the conclusion that the major shortcoming of <u>Fans and Systems</u> was the lack of system effect data for axial flow fans.

The technical committee reviewed any new, available literature on system effect factors for axial as well as centrifugal fans. Upon completion of the literature review, the committee proposed to the AMCA Board of Directors that an AMCA research project would provide the most definitive data possible on system effect factors for axial flow fans. The AMCA research project was approved in October 1986.

## 2. THE PROPOSAL

A few words are necessary in describing the technical committees philosophy in preparing the test program.

The committee chose to limit the program scope to specific system effects of primary interest to the AMCA membership. A limited test program was recommended to maintain a manageable amount of data and minimize the testing costs. Test results were monitored for validity immediately upon completion of a related series of tests, i.e.; if the test results were questionable for any reason, a check test could be undertaken while the test setup was still in place.

Axial fan test results were analyzed not only for the system effect factor as defined in <u>Fans and Systems</u>, but also for the effect of the system on fan power, fan stability and sound level. A system effect factor (SEF) is defined as a pressure loss which recognizes the effect of fan inlet restrictions, fan outlet restrictions or other conditions influencing fan performance when installed in the system.

The test program was established to determine system effect factors for varying lengths of outlet duct and the effects of two and four piece mitered elbows installed in close proximity to the fan inlet and outlet. A minimal amount of system effect information was also obtained on changes in the propeller hub/tip ratio and blade solidity.

The axial fans chosen for testing by the technical committee were a commercial grade of tubeaxial and vaneaxial fan. The tubeaxial fan selected for the test program has a peak total efficiency between 70 and 75 percent while the vaneaxial fan has a peak total efficiency between 72 and 78 percent. The test fan size was selected nominally as a 915 mm diameter propeller. Figure 1 provides dimensional and design specifics on the tubeaxial and vaneaxial fans tested. It is the responsibility of the user and fan manufacturer to determine whether the results of this testing are applicable to his axial fan design.

A description of the elbows tested is as follows:
  (a) two piece, round, mitered, 90 degree with a radius to diameter (r/d) ratio of 0.60 (approximate).
  (b) four piece, round, mitered, 90 degree with a radius to diameter (r/d) ratio of 1.50.

A total of forty-four (44) tests were proposed establishing pressure loss tests of fittings and selected system effect factors. All tests were run in the AMCA laboratory. The fan testing was done in accordance with AMCA Standard 210-85, Laboratory Methods of Testing Fans for Rating, either Figure 12, Outlet Chamber Setup - Multiple Nozzles in Chamber or Figure 15, Inlet Chamber Setup - Multiple Nozzles in Chamber. See Appendix No. 1 for a summary of the proposed testing.

In analyzing fan test results, it is a natural tendency for engineers to account for the smallest of differences between two fan test curves. The AMCA Fan Application Manual technical committee felt that a more macroscopic viewpoint was required with regard to system effect factors lest the committee get delayed by minor anomalies. System effect factors were initially established to quantify for users those system configurations that could significantly alter the fan output. The technical committee utilized only data where clear cut trends were indicated. Engineering judgment and experience were utilized to some extent where partial testing was done to minimize test costs, i.e.; blade solidity, hub/tip ratio and four piece mitered elbow.

## 3. SYSTEM EFFECT FACTOR CALCULATIONS

The system effect factor (SEF) for a given flow rate as determined by test is defined as follows:

$$sef = P_b - P_{sef} - \Delta P_{el} \pm fr$$

Where: $sef$ — System effect factor, pascals (See Figure 2)
$P_b$ — Fan static pressure without sef, pascals
$P_{sef}$ — Fan static pressure with sef, pascals
$\Delta P_{el}$ — Pressure loss of elbow, pascals
$fr$ — Friction pressure loss differences between tests with and without sef (as determined from ACMA Standard 210-85, Fig. 19, Friction Factors For Ducts)

System effect curves for all tests were determined at three points of operation, namely; near the peak of the pressure curve, the center and right of the recommended fan selection range.

The test setup to determine the pressure loss in the elbows is shown in Figure 3 along with a curve of the resultant pressure losses for the two and four piece elbows tested.

## 4. SYSTEM EFFECT FACTORS

### 4.1 Length of outlet duct

Fans intended for use with duct systems are usually tested with an outlet duct in place. The performance of fans may be affected when little or no outlet duct is used in application. The effect of varying lengths of outlet duct were obtained on the tubeaxial and vaneaxial fan tested in the AMCA research program. A summary of testing accomplished is shown in Tables 1 and 2 below.

Table 1   Tubeaxial fan (See Figure 4)

| Test No. | No. of Blades | H/T Ratio | Lgth. of Duct (D) |
|----------|---------------|-----------|-------------------|
| AT1 | 6 | .33 | 2743mm |
| AT2 | 6 | .25 | 2743mm |
| AT3 | 4 | .25 | 2743mm |
| AT4 | 8 | .25 | 2743mm |
| AT5 | 6 | .25 | 914mm |
| AT6 | 6 | .25 | 457mm |
| AT7 | 6 | .25 | 0mm |

Table 2   Vaneaxial fan (See Figure 4)

---

| Test No. | No. of Blades | H/T Ratio | Lgth. of Duct (D) |
|----------|---------------|-----------|-------------------|
| AV1 | 5 | .61 | 2743mm |
| AV2 | 9 | .61 | 2743mm |
| AV3 | 7 | .61 | 2743mm |
| AV4 | 7 | .61 | 914mm |
| AV5 | 7 | .61 | 457mm |
| AV6 | 7 | .61 | 0mm |

---

The general test arrangement and comparison curves for varying outlet duct length on both the tubeaxial and vaneaxial fans is shown in Figure 4. Duct lengths tested were 2743mm (3D), 914mm(1D), 457mm(0.5D) and 0mm(0.0D). Other system effect factor data was obtained with varying hub/tip ratio and blade solidity.

The test results for tubeaxial fans suggest that no system effect factor need be applied when outlet ducts are not installed. No significant differences were noticed in fan stability with and without outlet ductwork. Sound data was not recorded. Throughout all of this testing, sound data was only recorded where significant differences were discernible by the ear. The absence of a system effect factor for tubeaxial fans with little or no outlet duct is consistent with testing reported in AMCA paper 1950-86-A6 titled Discharge Diffuser Effect On Performance – Axial Fans by Lynne Galbraith of American Coolair, Inc..

One of the difficulties encountered in the analysis of system effect test data occurs at or near the peak pressure where installation parameters can sometime affect the onset of stall. In the case of the tubeaxial fan and reduced outlet duct length, the result was a positive system effect factor near the peak pressure. A positive system effect factor is one in which the catalog flow-pressure is exceeded as a result of the fan installation. For the purpose of the revised edition of AMCA Publication 201, Fans and Systems, no system effect factor will be utilized for tubeaxial fans with little or no outlet ducts.

The test results for vaneaxial fans with reduced lengths of outlet duct show that no system effect factor is significant until the duct length is reduced below one diameter. The revision to AMCA Publication 201, Fans and Systems, will recommend that 100 percent effective duct length be one duct diameter for vaneaxial fans, i.e.; no system effect factor is to be applied for outlet duct lengths of one diameter or greater.

See Table 3 below for system effect curves on vaneaxial fans with reduced lengths of outlet duct.

Table 3   System effect curves – vaneaxial fans with minimal outlet duct length (Determine SEF using Figure 6)

---

| Fan Type | 0.0D | 0.5D | 1.0D | 3.0D |
|----------|------|------|------|------|
| Vaneaxial | U | W | — | — |

---

No system effect for power is recommended because of the small power differences. The use of the catalogued power for the desired flow and pressure including the system effect factor will provide a suitable motor selection. No significant differences in fan stability or fan sound were observed on either the tubeaxial or vaneaxial fans with reduced lengths of outlet duct.

4.2   Outlet elbows

Obstructions located at or near a fan outlet, before a uniform velocity is established by controlled diffusion, may result in reduced fan performance, i.e.; a system effect has occurred. Testing was undertaken on a tubeaxial and vaneaxial fan to determine the effect that elbows in close proximity to the fan outlet have on fan performance. A summary of the testing accomplished is shown in Tables 4 and 5 below.

Table 4   Tubeaxial   fan  with   outlet
            elbows  (See Figure 5)

| Test No. | Elbow | A | B | H/T Ratio | No. of Blades |
|---|---|---|---|---|---|
| AT8 | 2 Pc. | 0mm | 1829mm | .25 | 6 |
| AT9 | 2 Pc. | 457mm | 1829mm | .25 | 6 |
| AT10 | 2 Pc. | 914mm | 1829mm | .25 | 6 |
| AT11 | 2 Pc. | 2743mm | 1829mm | .25 | 6 |
| AT12 | 2 Pc. | 0mm | 1829mm | .33 | 6 |
| AT13 | 2 Pc. | 0mm | 1829mm | .25 | 4 |
| AT14 | 2 Pc. | 0mm | 1829mm | .25 | 8 |
| AT15 | 4 Pc. | 0mm | 1829mm | .25 | 6 |
| AT16 | 4 Pc. | 0mm | 0mm | .25 | 6 |

Table 5  Vaneaxial fan with outlet
           elbows  (See Figure 5)

| Test No. | Elbow | A | B | H/T Ratio | No. of Blades |
|---|---|---|---|---|---|
| AV7 | 2 Pc. | 0mm | 1829mm | .61 | 7 |
| AV8 | 2 Pc. | 457mm | 1829mm | .61 | 7 |
| AV9 | 2 Pc. | 914mm | 1829mm | .61 | 7 |
| AV10 | 2 Pc. | 2743mm | 1829mm | .61 | 7 |
| AV11 | 2 Pc. | 0mm | 1829mm | .61 | 9 |
| AV12 | 2 Pc. | 0mm | 1829mm | .61 | 5 |
| AV13 | 4 Pc. | 0mm | 1829mm | .61 | 7 |

The general test arrangement and comparison curves for two and four piece mitered outlet elbows in close proximity to the fan for both tubeaxial and vaneaxial fans is shown in Figure 5. The lengths of duct tested between the fan outlet and elbow were 2743mm(3D), 914mm(1D), 457mm(0.5D) and 0mm(OD). Other system effect factor data was obtained with varying hub/tip ratio and blade solidity.

The system effect factor for a two piece mitered outlet elbow on a tubeaxial fan is negligible regardless of the length of ductwork between the fan outlet and the elbow. The changes in power were correspondingly small. Similar test results were obtained on a four piece mitered outlet elbow mounted directly on the tubeaxial fan outlet. Changes in hub/tip ratio (0.25 and 0.33) and blade number (4, 6 and 8) had little affect on the performance of a tubeaxial fan with an outlet elbow.

The revised version of AMCA Publication 201, _Fans and Systems_, will not apply a system effect factor to outlet elbows at or in close proximity to tubeaxial fans. Catalog performance for tubeaxial fans with outlet elbows should be reasonably maintained provided all system losses are properly accounted for including the pressure loss of the elbow.

The AMCA technical committee fully expected a system effect factor for tubeaxial fans when utilized without an outlet duct or with an elbow in close proximity to the fan outlet. The test data is believed to be reliable because of the trends established in this series of tests and the strict adherence to test procedures per AMCA Standard 210-85. It is surmised by the committee that the spiraling velocity profile exiting the tubeaxial fan outlet maintains its profile throughout the given duct condition and does not affect the fan output as with the straight through flow of a centrifugal or vaneaxial fan.

Small system effect factors were obtained for elbows in close proximity to a vaneaxial fan outlet. The system effect curves for 90 degree mitered elbows near to a vaneaxial fan outlet are given in Table 6 below.

Table 6   System  effect  curves  –
           vaneaxial  fans  with  90
           degree  mitered  outlet
           elbows (Determine SEF using
           Figure 6)

| Duct Length | 0.0D | 0.5D | 1.0D | 3.0D |
|---|---|---|---|---|
| 2 Pc. | U | V | W | ___ |
| 4 Pc. | W | ___ | ___ | ___ |

The number of blades in the propeller had no appreciable affect on the system effect factors for vaneaxial fans with outlet elbows. No fan instability or increased sound levels were observed.

© IMechE 1990 C401/006

## 4.3 Inlet elbows

Distorted velocity profiles entering a fan inlet will reduce the usable output of a fan. Elbows located at or in close proximity to a fan inlet are a common source of velocity distortion at fan inlets. A tubeaxial and vaneaxial fan were tested to quantify the effect that inlet elbows in close proximity to the fan inlet have on the fan performance. A summary of the testing accomplished is shown in Tables 7 and 8 below.

Table 7 Tubeaxial fan with inlet elbows (See Figure 7)

| Test No. | Elbow | A | H/T Ratio | No. of Blades |
|----------|-------|-----|-----------|----------------|
| AT17 | 2 Pc. | 0mm | .25 | 6 |
| AT18 | 2 Pc. | 457mm | .25 | 6 |
| AT19 | 2 Pc. | 914mm | .25 | 6 |
| AT20 | 2 Pc. | 2743mm | .25 | 6 |
| AT21 | 2 Pc. | 0mm | .33 | 6 |
| AT22 | 2 Pc. | 0mm | .25 | 4 |
| AT23 | 2 Pc. | 0mm | .25 | 8 |
| AT24 | 4 Pc. | 0mm | .25 | 6 |

Table 8 Vaneaxial fan with inlet elbows (See Figure 7)

| Test No. | Elbow | A | H/T Ratio | No. of Blades |
|----------|-------|-----|-----------|----------------|
| AV21 | None | ___ | .61 | 7 |
| AV22 | 2 Pc. | 0mm | .61 | 7 |
| AV23 | 2 Pc. | 457mm | .61 | 7 |
| AV24 | 2 Pc. | 914mm | .61 | 7 |
| AV25 | 2 Pc. | 2743mm | .61 | 7 |

The general test arrangement and comparison curves for mitered inlet elbows in close proximity to tubeaxial and vaneaxial fans is shown in Figure 7. The lengths of duct tested between the fan inlet and elbow were 2743mm(3D), 914mm(ID), 457mm(0.5D) and 0mm(OD). Some system effect factor data was obtained on tubeaxial fans with varying hub/tip ratio and blade solidity.

System effect factors are shown in Table 9 below for tubeaxial and vaneaxial fans with two and four piece mitered elbows at or in close proximity to the fan inlets. Another variable tested was blade solidity. The number of blades did not have a significant effect on inlet elbow system effect factors.

Table 9 System effect curves - tubeaxial and vaneaxial fans with 90 degrees mitered inlet elbows (Determine SEF by using Figure 6)

| Fan Type | H/T Ratio | Elbow | 0.0D | 0.5D | 1.0D | 3.0D |
|----------|-----------|-------|------|------|------|------|
| Tube axial | .25 | 2 Pc. | U | V | W | ___ |
| Tube axial | .25 | 4 Pc. | X | ___ | ___ | ___ |
| Tube axial | .33 | 2 Pc. | V | W | X | ___ |
| Vane axial | .61 | 2 Pc. | Q-R | Q-R | S-T | T-U |

A word of caution is required with the use of inlet elbows in close proximity to fan inlets. Other than the loss of usable fan performance, instability in fan operation may occur as evidenced by an increase in pressure fluctuations and fan sound level. Fan instability may result in serious structural damage to the fan. Serious fan instabilities with inlet elbows in close proximity to the vaneaxial fan were experienced during the AMCA research project. The fan instabilities were not experienced with the testing of the tubeaxial fan.

Pressure fluctuations on the manometers for vaneaxial fans with inlet elbows approached ten times the magnitude of fluctuations for the fan with a fully developed velocity profile at the inlet. At the r/min of the vaneaxial test fan, the total pressure peaked at approximately 560 pascals. Pressure fluctuations at this point of fan operation approached 75 pascals peak-to-peak. Significant increases in sound level were experienced with inlet elbows on the vaneaxial fan as shown in Figures 8 and 9. The increase in fan sound level is most predominant in the lower frequency bands.

**It is strongly advised that inlet elbows be installed an absolute minimum of three diameters away from any axial or centrifugal fan inlet.**

## 5. SUMMARY

The AMCA Fan Application Manual Review Committee believes that the testing undertaken on this research project has provided valuable information on system effect factors for tubeaxial and vaneaxial fans. Basic fan design information on the test fans is provided to allow the user to make conclusions about the applicability of the data to his fan design. The results of this AMCA research project are to be utilized in the upcoming revision to AMCA Publication 201, Fans and Systems.

As stated previously, the system effect factors for the research project were checked at three points of fan operation spanning the normal selection range for tubeaxial and vaneaxial fans. Except for those cases where the onset of stall was affected by the duct system, the system effect curve was reasonably constant regardless of the point of operation.

The use of system effect factors for power were reviewed in the analysis of data for this project. The power data confirmed the assumption made in the current edition of AMCA Publication 201, Fans and Systems, i.e.; when all of the applicable system effect factors have been added to the calculated system pressure losses, the power given in the fan manufacturers catalog for that point of operation may be used without adjustment.

APPENDIX NO. 1  TURBAXIAL FAN SEF TESTING
Page 1 of 2

| TEST NO. | BLADE SOLIDITY 4 | 6 | β | H/T RATIO .25 | .33 | DUCT LENGTHS 0 | 18" | 36" | 108" | DUCT CONF. INLET | OUTLET | ELBOWS 2 PC. | 4 PC. | AMCA FIG. #12 | #15 | 72" INLET/ OUTLET DUCT | BELL MOUTH | BASE TEST | REMARKS |
|---|---|---|---|---|---|---|---|---|---|---|---|---|---|---|---|---|---|---|---|
| 1 |  | X |  |  | X |  |  |  |  |  | X |  |  |  | X |  | X | 2 |  |
| 2 |  | X |  | X |  |  |  |  | X |  | X |  |  |  | X |  | X |  | H/T Ratio Effect |
| 3 | X |  |  | X |  |  |  |  | X |  | X |  |  |  | X |  | X | 2 |  |
| 4 |  |  | X | X |  |  |  |  | X |  | X |  |  |  | X |  | X | 2 | Blade Solidity Effect |
| 5 |  | X |  | X |  |  | X |  |  |  | X |  |  |  | X |  | X | 2 |  |
| 6 |  | X |  | X |  |  |  | X |  |  | X |  |  |  | X |  | X | 2 |  |
| 7 |  | X |  | X |  | X |  |  |  |  | X |  |  |  | X |  | X | 2 | Evaluate Outlet Duct Length SEF |
| 8 |  | X |  | X |  | X |  |  |  |  | X | X |  |  | X | X | X | 2 |  |
| 9 |  | X |  | X |  | X |  |  |  |  | X | X |  |  | X | X | X | 2 |  |
| 10 |  | X |  | X |  | X |  |  |  |  | X | X |  |  | X | X | X | 2 |  |
| 11 |  | X |  | X |  | X |  |  |  |  | X | X |  |  | X | X | X | 2 | Evaluate 2 Pc. Disch. Elbow SEF |
| 12 |  | X |  |  | X | X |  |  |  |  | X | X |  |  | X | X | X | 1 | H/T Ratio Effect on Disch Elbow SEF |
| 13 | X |  |  | X |  | X |  |  |  |  | X | X |  |  | X | X | X | 3 |  |
| 14 |  |  | X | X |  | X |  |  |  |  | X | X |  |  | X | X | X | 4 | Bld. Sol. Effect on Disch. Elbow SEF |
| 15 |  | X |  | X |  | X |  |  |  |  | X |  | X |  | X | X | X | 2 | Evaluate 4 Pc. Disch. Elbow SEF |
| 16 |  | X |  | X |  |  |  |  | X |  | X |  | X |  | X | X | X | 15 | Effect of Outlet Duct on Elbow SEF |
| 17 |  | X |  | X |  | X |  |  |  | X |  | X |  | X | X |  | X | 2 |  |
| 18 |  | X |  | X |  | X |  |  |  | X |  | X |  | X | X |  | X | 2 |  |
| 19 |  | X |  | X |  |  |  | X |  | X |  | X |  | X | X |  | X | 2 |  |
| 20 |  | X |  | X |  | X |  |  |  | X |  | X |  | X | X |  | X | 2 | Evaluate 2 Pc. Inlet Elbow SEF |
| 21 |  | X |  | X |  |  |  |  | X | X |  | X |  | X | X |  | X | 1 | H/T Ratio Effect on Inlet Elbow SEF |
| 22 | X |  |  | X |  | X |  |  |  | X |  | X |  | X | X |  | X | 3 |  |
| 23 |  |  | X | X |  | X |  |  |  | X |  | X |  | X | X |  | X | 4 | Blade Sol. Effect on Inlet Elbow SEF |
| 24 |  | X |  | X |  | X |  |  |  | X |  |  | X | X | X |  | X | 2 | Evaluate 4 Pc. Inlet Elbow Effect |

APPENDIX NO. 1 VANEAXIAL SEF TESTING
Page 2 of 2

| TEST NO. | BLADE SOLIDITY | | | DUCT LENGTHS | | | | DUCT. CONF. | | ELBOWS | | ANCA FIG. | | 72" INLET/OUTLET DUCT | BELL MOUTH | BASE TEST | REMARKS |
|---|---|---|---|---|---|---|---|---|---|---|---|---|---|---|---|---|---|
| | A-5 | B-6 | C-7 | 0 | 18" | 36" | 108" | INLET | OUTLET | 2 PC. | 4 PC. | 012 | 015 | | | | |
| 1 | X | | | | | | X | | X | | | | X | | X | | |
| 2 | | X | | | | | X | | X | | | | X | | X | | |
| 3 | | | X | | | | X | | X | | | | X | | X | 3 | Blade Solidity Effect |
| 4 | | | X | | | X | | | X | | | | X | X | X | 3 | |
| 5 | | | X | | X | | | | X | | | | X | X | X | 3 | Evaluate Outlet Duct Length SEF |
| 6 | | | X | X | | | | | X | X | | | X | X | X | 3 | |
| 7 | | | X | X | | | | | X | X | | | X | X | X | 3 | |
| 8 | | | X | | X | | | | X | X | | | X | X | X | 3 | |
| 9 | | | X | | | | X | | X | X | | | X | X | X | 3 | Evaluate 2 Pc. Discharge Elbow SEF |
| 10 | | | X | X | | | | | X | X | | | X | X | X | 3 | |
| 11 | | X | | X | | | | | X | X | | | X | X | X | 2 | Blade Sol. Effect on Discharge Elbow SEF |
| 12 | X | | | X | | | | | X | X | | | X | X | X | 1 | |
| 13 | | | X | X | | | | | X | | X | | X | X | X | 3 | Evaluate 4 Pc. Discharge Elbow SEF |
| 14 | | | X | | X | | | X | | X | | X | | X | X | 3 | |
| 15 | | | X | X | | | | X | | X | | X | | X | X | 3 | Evaluate 2 Pc. Inlet Elbow SEF |
| 16 | | | X | | | X | | X | | X | | X | | X | X | 3 | |
| 17 | | | X | | | | X | X | | X | | X | | X | X | 3 | |
| 18 | | X | | X | | | | X | | X | | X | | | X | 2 | Blade Sol. Effect on Inlet Elbow SEF |
| 19 | X | | | X | | | | X | | | | X | | X | X | 1 | |
| 20 | | | X | X | | | | X | | | X | X | | X | X | 3 | Evaluate 4 Pc. Inlet Elbow SEF |

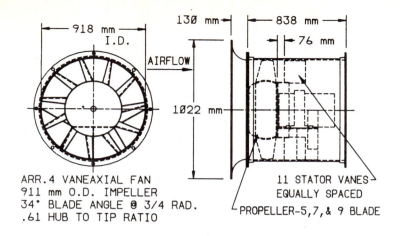

ARR.4 VANEAXIAL FAN
911 mm O.D. IMPELLER
34° BLADE ANGLE @ 3/4 RAD.
.61 HUB TO TIP RATIO

11 STATOR VANES
EQUALLY SPACED

PROPELLER-5,7,& 9 BLADE

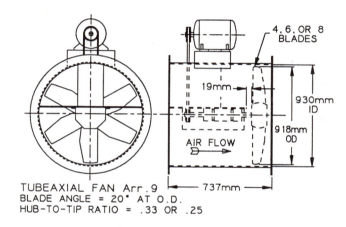

TUBEAXIAL FAN Arr.9
BLADE ANGLE = 20° AT O.D.
HUB-TO-TIP RATIO = .33 OR .25

Fig 1     Vane axial and tube axial fan dimensional information

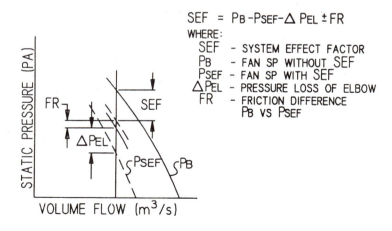

$$SEF = P_B - P_{SEF} - \Delta P_{EL} \pm FR$$

WHERE:
SEF  – SYSTEM EFFECT FACTOR
$P_B$    – FAN SP WITHOUT SEF
$P_{SEF}$ – FAN SP WITH SEF
$\Delta P_{EL}$ – PRESSURE LOSS OF ELBOW
FR   – FRICTION DIFFERENCE
       $P_B$ vs $P_{SEF}$

Fig 2     System effect factor

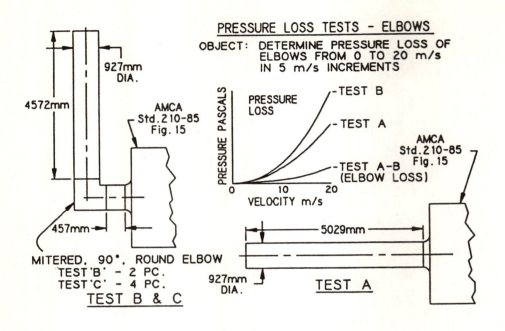

PRESSURE LOSS TESTS - ELBOWS

OBJECT: DETERMINE PRESSURE LOSS OF
ELBOWS FROM 0 TO 20 m/s
IN 5 m/s INCREMENTS

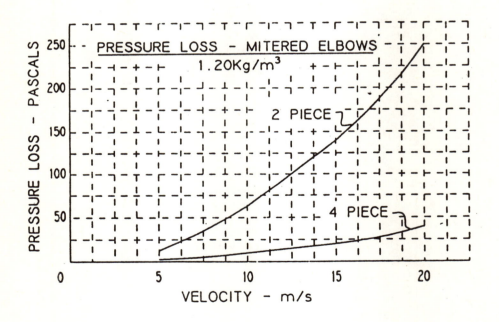

Fig 3    Elbow pressure losses

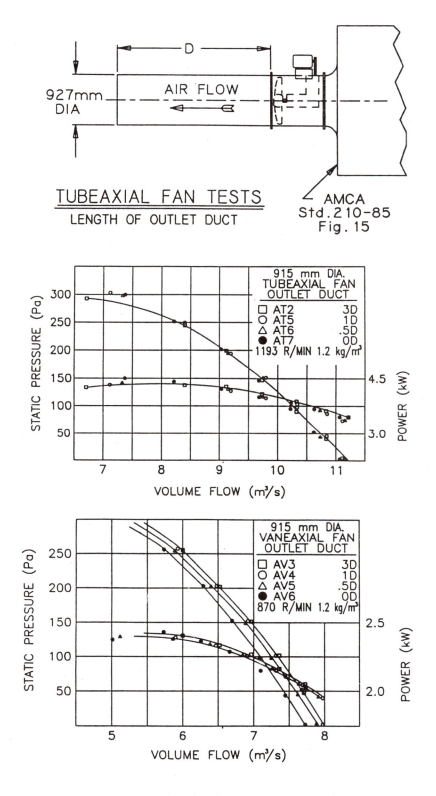

Fig 4     Outlet duct length test arrangements

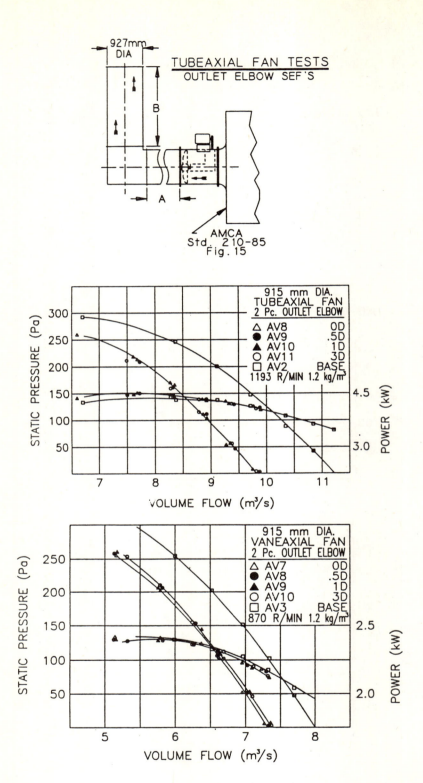

Fig 5    Outlet elbow test arrangements

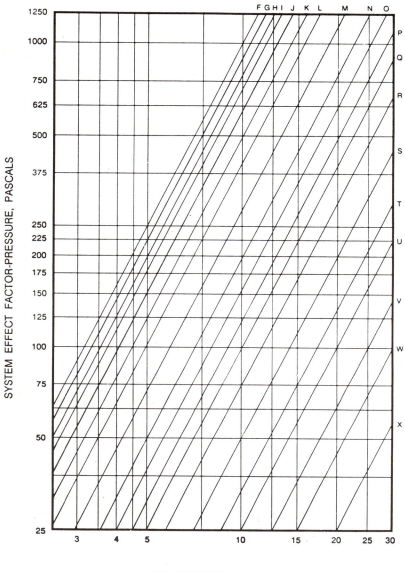

AIR VELOCITY, m/s

Air Density = 1.2 Kg/m³

Fig 6     System effect curves

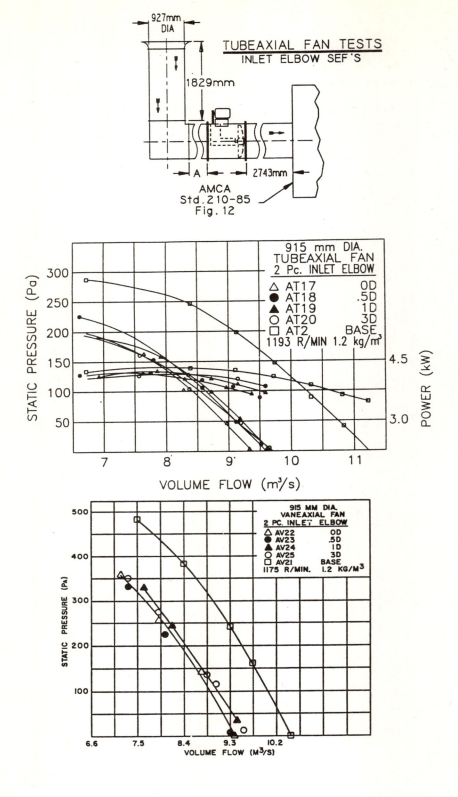

Fig 7    Inlet elbow test arrangements

# MEASURED INLET SOUND POWER

## VANEAXIAL FAN SYSTEM EFFECT
### TEST 6631-S1, S2, S3

**FAN ONLY**   **2 PC ELBOW ON INLET**   **2 PC ELBOW 3 De**

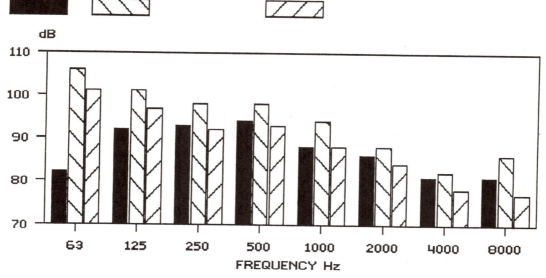

(BLADE PASS 135 Hz)

Fig 8    Inlet sound power

| Hz | S1 | S2 | S3 |
|---|---|---|---|
| 50 | 79 dB | 106 dB | 101 dB |
| 63 | 73 | 86 | 82 |
| 80 | 78 | 99 | 88 |
| 100 | 77 | 93 | 90 |
| 125 | 91 | 99 | 95 |
| 160 | 83 | 89 | 91 |
| 200 | 86 | 92 | 80 |
| 250 | 91 | 95 | 91 |
| 315 | 86 | 93 | 86 |
| 400 | 92 | 93 | 89 |
| 500 | 88 | 94 | 88 |
| 630 | 87 | 92 | 87 |
| 800 | 85 | 91 | 85 |
| 1.0K | 83 | 88 | 82 |
| 1.3K | 83 | 86 | 82 |
| 1.6K | 83 | 84 | 80 |
| 2.0K | 82 | 83 | 79 |
| 2.5K | 79 | 81 | 77 |
| 3.2K | 76 | 78 | 74 |
| 4.0K | 76 | 77 | 73 |
| 5.0K | 75 | 76 | 71 |
| 6.3K | 77 | 76 | 71 |
| 8.0K | 77 | 79 | 70 |
| 10.00K | 75 | 85 | 73 |

S1  =  Fan with Inlet Bell

S2  =  Fan with 2 piece elbow adjacent

S3  =  Fan with 2 piece elbow 3 D away

ONE THIRD OCTAVE BAND CHART

BLADE PASS FREQUENCY 135 Hz

$L_{wmi}$ MEASURED INLET SOUND POWER

Fig 9    Inlet sound power — vane axial fan

# Installation effects in fan systems

**J A RIERA-UBIERGO** and **F CHARBONNELLE**
Department of Aerodynamic, Acoustic and Air Diffusion, Centre Technique des Industries Aerauliques et Thermiques (CETIAT), Orsay, France

SYNOPSIS An experimental investigation has been carried out on the influence of duct fittings on a centrifugal and on an axial fan performance. This paper presents a qualitative and quantitative analysis from 11 different upstream configurations.

## I - INTRODUCTION

Inlet and outlet duct configuration significantly affects fan performance in terms of pressure generation, power consumption and efficiency. For that reason, it is important for a duct system designer to know both qualitatively and quantitatively the influence of his installation on a fan so that he can optimize the duct fittings and eventualy choose a suitable fan.

Due to the lack of experimental data and to the difficulty of undertaking theoretical investigations, there is still a need for more reliable information about the effect of duct fittings on fan aerodynamic performance.

This paper presents some results from preliminary tests concerning the influence of several inlet duct configurations on the performance of a centrifugal and an axial fan.

## II - DESCRIPTION

The centrifugal fan is a 64 bladed forward curved centrifugal fan. The fan has a 350 mm inlet diameter and operates at a 1000 r.p.m rotational speed. The geometrical characteristics of the impeller and scroll are given in Figure 1a. The axial fan is a 400 mm tip diameter, 6 bladed tubeaxial fan running at a 2900 r.p.m rotational speed. Fig. 1a presents the principal dimensions.

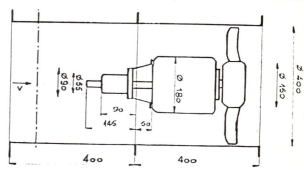

Fig 1b    Axial fan

## III - DESCRIPTION OF INLET DUCTS COMPONENTS

Each fan has been tested within eleven different upstream duct configurations : one configuration without any ducts that we shall call configuration number 0 plus ten other configurations (from number 1 to number 10) made up from two 400 mm diameter duct components.

The first component (called element S) is a 10 diameter long straight cylindrical duct which is composed of (Fig. 2a) :

- a 3 diameters long straight duct in the downstream part of which, is installed an AMCA flow straightener. This flow straightener is a honeycomb type staightener made up from square cells of 75 mm width,

- a 7 diameters long straight duct.

The second component (called element E) is a 90° segmented elbow of radius ratio (inside radius/inlet diameter) equal to one (Fig. 2b). This elbow is made up from 4 straight sections and has no turning vanes.

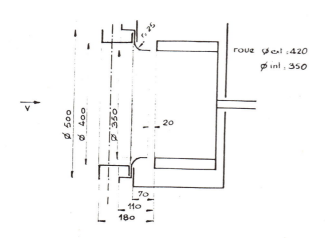

Fig 1a    Centrifugal fan

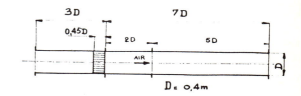

Fig 2a    Element S

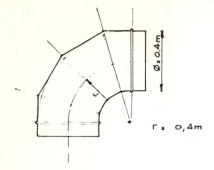

Fig 2b    Element E

All the inlet duct configurations are shown in Fig.3 for the centrifugal fan as an example. These configurations are the same for both fans and are summarized in the following table.

| Inlet conf. N° | N° of element S | N° of element E | Fittings from fan inlet |
|---|---|---|---|
| 0 | 0 | 0 | - |
| 1 | 1 | 0 | S |
| 2 | 1 | 2 | S + E + E |
| 3 | 1 | 1 | S + E |
| 4 | 1 | 1 | E + S |
| 5 | 1 | 2 | E + E + S |
| 6 | 1 | 2 | E + E + S |
| 7 | 1 | 2 | E + E + S |
| 8 | 1 | 1 | E + S |
| 9 | 1 | 2 | S + E + E |
| 10 | 1 | 2 | S + E + E |

Table 1

## IV - DESCRIPTION OF EXPERIMENTAL TESTS

Each fan complete with fittings has been tested according to a procedure specified in the French standard NF X 10-200, (reference 1) in which the outlet of each fan was connected to a standard installation of type B.

This installation comprises three parts ; from fan outlet we have respectively :

- A common part composed of straight cylindrical ducts and a straightener.
- A 7 degree diverging diffuser.
- A 1.2 m diameter plenum incorporating a honeycomb-type straightener and a measuring section for the pressure.

For each configuration, sets of measurements were made at 6 operating points in order to determine :

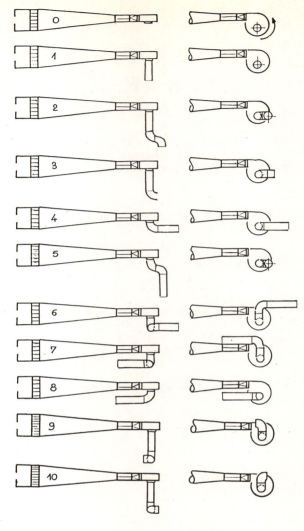

Fig 3    Inlet duct fittings

- the fan outlet total pressure rise versus flowrate
- input power versus flowrate
- total efficiency versus flowrate.

A total of 22 tests have been carried out.

## V - RESULTS AND REMARKS

We can divide the remarks into two groups :

- qualitative remarks
- quantitative remarks.

### V-1 - QUALITATIVE REMARKS

If we examine the pressure, power and efficiency characteristic curves obtained from each test without making any correction for component pressure loss we can derive some interesting qualitative conclusions about the different inlet fittings.

The effect of the installation can be examined :

    1 - with respect to the power consumption
and 2 - With respect to pressure rise and efficiency.

## V-1-1 - <u>POWER CONSUMPTION</u>

The power consumption versus flowrate characteristic for each are shown in Fig. 4.

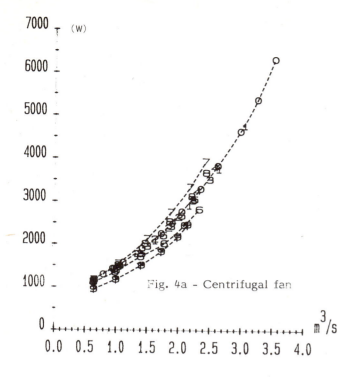

Fig. 4a – Centrifugal fan

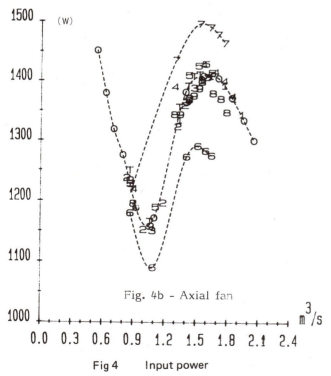

Fig. 4b – Axial fan

Fig 4    Input power

Two groups of inlet configurations can be distinguished.

### GROUP A

This group comprises inlet configurations in which the outlet part (linked to the fan inlet) is elements S (configurations 1, 2, 3, 9 and 10).

Concerning configuration 1, 2 and 3 all the power characteristics are the same as for the free inlet configuration 0. It means that those configurations do not affect the power consumption.

For the centrifugal fans, configurations 9 and 10 tend to decrease the power consumption.

To conclude, this combination of inlet fittings is beneficial in terms of power consumption because the air is entering axially and uniformly.

### GROUP B

This group comprises configurations in which the outlet part (linked to the fan inlet) is composed of one or more elbows (element E) (configurations 4, 5, 6, 7 and 8).

- For both fans, configurations 6 and 7 result in the greatest differences in power consumption. These installations differ from each other in the way that configuration 6 is a mirror image of configuration 7. It can be seen that their power consumption versus flowrate characteristics vary symmetrically about the one at free inlet (configuration 0). This can be explained by the fact that a combination of two 90° elbows in two different planes induces a strong swirling flow around the rotation axis at the fan inlet. In configuration 6, the fan inlet flow presents a swirl in the same direction as the impeller rotation : it results in a decrease in power. In configuration 7, due to a counter rotating vortex at the inlet it results in an increase in power.

- For configuration 4 and 5, the power characteristics are similar whatever the fan. In the case of the centrifugal fan, the characteristics are similar to the free inlet condition configuration 0. In the case of the axial fan the power consumption is slightly greater than for the free inlet condition configuration. The common features of these configurations are for the ducts to be in the same plane with a 90° elbow oriented in the same direction at the fan inlet. It appears that this type of configuration induces a small counter rotating inlet swirl.

- As for configurations 6 and 7, configuration 8 is a mirror image of configuration 4. It can be seen that in the case of the axial fan the power characteristics of configurations 4 and 8 are symmetrical about the one at free inlet condition.

In conclusion, the presence of one or several elbows at a fan inlet induces a swirl which can either reduce or increase the fan's power consumption according to the direction of rotation of the induced vortex relative to the direction of fan rotation.

## V-1-2 - <u>PRESSURE RISE AND EFFICIENCY</u>

The total pressure rise versus flowrate characteristic for each fan are shown in Fig. 5. The efficiency characteristic curves are shown in Fig. 6.

In the following section installation configurations which have the same number and kind of elements are compared. It is assumed that the overall pressure loss induced by the duct fittings is independant of the combination of the elements.

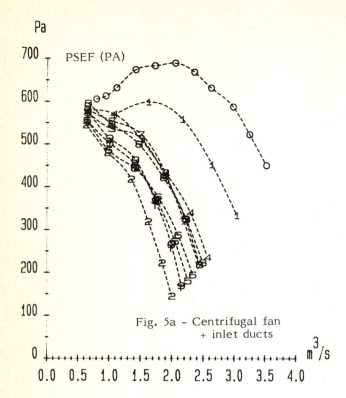

Fig. 5a - Centrifugal fan
+ inlet ducts

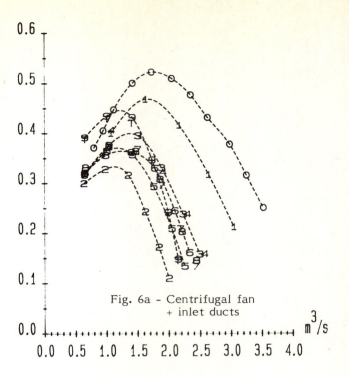

Fig. 6a - Centrifugal fan
+ inlet ducts

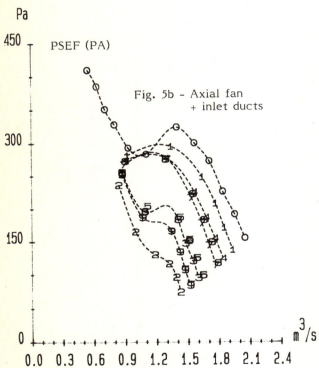

PSEF (PA)

Fig. 5b - Axial fan
+ inlet ducts

**Fig 5    Total pressure versus flowrate**

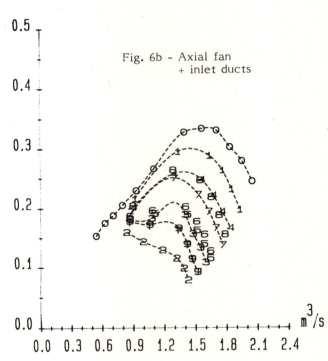

Fig. 6b - Axial fan
+ inlet ducts

**Fig 6    Total efficiency versus flowrate**

Three groups of combinations can be distinguished :

A) S + E + E (in different planes) : combination of one element S with two element E in different planes.

This group comprises configurations 6, 7, 9 and 10.

From this group, it results that (in terms of pressure) configurations 9 and 10 are the most sensitive to system effects. It will be noticed

that whatever the fan, all the characteristics (pressure, power, efficiency versus flowrate) are similar for configuration 9 and 10. This tends to prove that the elbows orientation makes no difference on system effect when they are put on element S inlet. In the case of the axial fan, configuration 6 and 7 are more favorable than configuration 9 and 10 in terms of pressure and efficiency.

In the case of the centrifugal fan, configurations 6 and 7 result in higher pressures than configurations 9 and 10 but not in higher efficiencies.

The pressure characteristic of configuration 7 is higher than that of configuration 6 whatever the fan. In terms of efficiency, configuration 7 is similar to configuration 6 with the axial fan. This tends to prove that an inlet counter-rotating swirl is more favorable than an inlet swirl in the direction of the impeller rotation.

To conclude with this group of configurations it seems better to install the two 90° elbows close to the fan inlet rather than at a distance.

B) S + E + E (in the same plane) combination of one element S first two elements E in the same plane.

This group concerns configurations 2 and 5. Configuration 5 is largely less sensitive to system effect than configuration 2. This is true whatever the type of fan for both pressure and efficiency (despite a slight over-consumption with the axial fan). Here again, the results seem to prove that it is preferable to put the elbows directly at fan inlet.

C) E + S (combination of one element E with one element S)

This group comprises configuration 3, 4 and 8. The configuration 8 and 4 have very similar pressure characteristics, this is due to the fact that configuration 8 is a mirror image of configuration 4.
Nevertheless configuration 4 is more favorable than configuration 8 whatever the fan type in pressure as in efficiency.

In the case of the axial fan, configurations 4 and 8 are more favorable than configuration 3 in pressure and in efficiency. In the case of the centrifugal fan, configuration 3 is located between configuration 8 and 4 in pressure and efficiency.

To conclude with qualitative analysis of the inlet duct fitting configurations two important points appear :

1 - Whatever the fan type it is preferable (for pressure and efficiency) to induce a counter-swirl at fan inlet rather than swirl in the direction of the impeller rotation, despite a power consumption increase.

2 - System effect seems to be less pronounced when elbows are installed first at the fan inlet instead of straight ducts.

V-2 - QUANTITATIVE REMARKS

For a given flowrate we can define a "System Effect Factor" (SEF) as follows :

$$SEF = PB - PSEF - PL$$

where

SEF : System Effect Factor
PB : Fan total pressure without any inlet components
PSEF : Fan total pressure measured with components on the inlet
PL : Total pressure loss due to inlet duct fittings.

PB corresponds to the pressure obtained in configuration 0.
PSEF depends on the inlet fitting configuration type (Fig. 9)
For the determination of PL, the following loss coefficients were adopted :

- Element S loss coefficient : 0,5
- Element E loss coefficient : 0,4
- 2 elbows connected in the same plane : 0,8
- 2 elbows not connected in the same plane : 0,64.

All those coefficients are non-dimensionalized with respect to an inlet dynamic pressure based on uniform flow going through a 400 mm diameter cylindrical section.
Fig. 7 and 8 show the total pressure and efficiency corrected with total pressure loss due to inlet duct fittings.

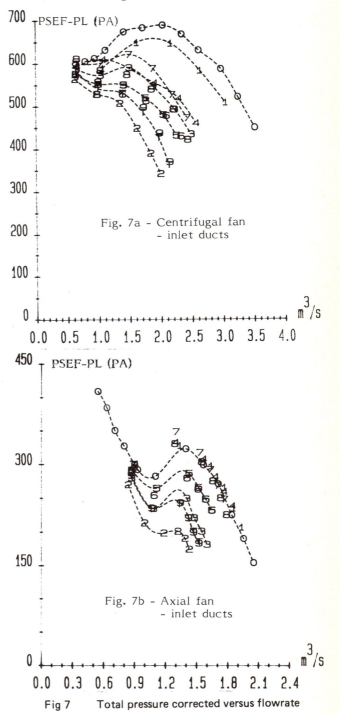

Fig. 7a - Centrifugal fan - inlet ducts

Fig. 7b - Axial fan - inlet ducts

Fig 7    Total pressure corrected versus flowrate

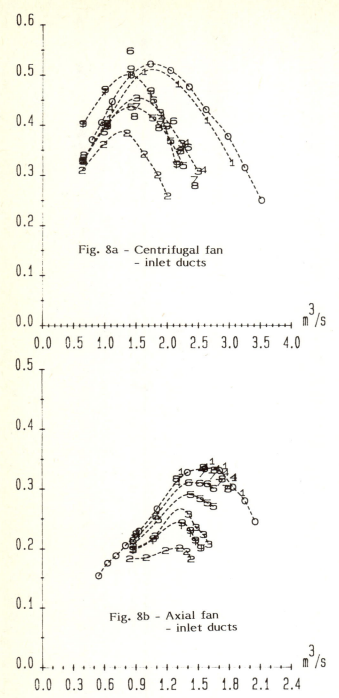

Fig. 8a – Centrifugal fan
– inlet ducts

Fig. 8b – Axial fan
– inlet ducts

Fig 8    Total efficiency corrected versus flowrate

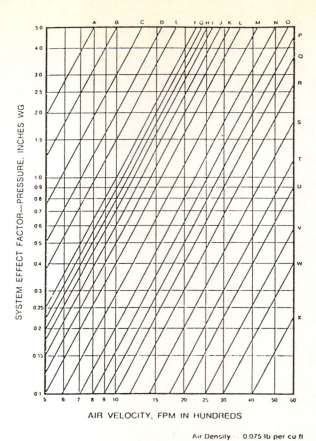

Air Density - 0.075 lb per cu ft

Fig 9    AMCA system effect curves

The system effect factors can then be converted to system effect curves according to the AMCA publication 201 - Fig. 9 - (reference 2).

This leads us to the table 2.

Looking at table 2, we can make the following conclusions :

- The centrifugal fan is always more sensitive to system effect than the axial fan, whatever the inlet duct fittings.
- No system effects appear for configuration 1, 4 and 7 in the case of the axial fan : configuration 1 is even better than configuration 0. The presence of elbows inducing counter rotating swirl at fan inlet seems to reduce system effects.
- For both fans, configuration 2 has the highest system effect factor.
- For both fans, configuration 1 has the lowest system effect factor.

List of references :

1. AFNOR : NF X 10-200 (1971) : Industrial fans - Rules for performance testing of ducted free inlet and fully ducted fans using standardized airways - "Caisson reduit" method at the outlet.

2. AMCA : Publication 201 : Fan application manual - FANS AND SYSTEMS.

| Inlet conf. N° | PL coef. | Axial fan SEF | Centrifugal fan SEF |
|---|---|---|---|
| 1 | 0,5 | No SE | W |
| 2 | 1,3 | P | P |
| 3 | 0,9 | R - S | R - S |
| 4 | 0,9 | No SE | S - T |
| 5 | 1,3 | T | Q - R |
| 6 | 1,14 | T | Q - R |
| 7 | 1,14 | No SE | S - T |
| 8 | 0,9 | W - X | R - S |
| 9 | 1,14 | Q - R | P - Q |
| 10 | 1,14 | Q - R | P - Q |

TABLE 2

# Accuracy of in-duct fan sound power level determination over a range of standard test installations

**A N BOLTON**, BSc, **A J GRAY** and **E J MARGETTS**, MIOA
National Engineering Laboratory, East Kilbride, Glasgow

SYNOPSIS

In duct noise measurements on two fans were undertaken at six European laboratories to determine the accuracy with which sound power levels could be derived. Analysis of the data showed that the effect of variations in test rig geometry, within the limits prescribed in the Draft ISO Standard DIS 5136, resulted in a scatter which was small compared to other sources of uncertainty. A number of factors were noted which had significant influences on repeatability and the findings from the tests will be used in the preparation of the ISO Standard for fan in-duct noise measurement.

## 1 INTRODUCTION

It is known that the generation and transmission of fan noise is influenced by the ductwork connected to the fan inlet and outlet. Theoretical and experimental work by Baade (1) and others (2,3) has demonstrated that changes in sound power generation are caused by variations in duct configuration which can alter both the impedance loading on the fan and the aerodynamic flow conditions into and out of the fan. In order to predict the noise of a fan operating in an arbitrary installation it will be necessary to have information relating both to the installation in which the fan was initially tested and to the proposed installation.

Standardised installation configurations for determining fan sound power levels, have been developed by various national and international standards organisations such as DIN, AFNOR, BSI and ISO. These installations attempt to provide an anechoic termination to the measurement duct(s) connected to a fan so that the impedance load on the fan approaches the ideal, resistive value of unity. For practical reasons it is not possible to make perfect anechoic terminations and the standards recognise this by specifying maximum limits of reflection coefficient. Further, the standards permit, for a given fan diameter, a range of acceptable sizes of test duct. Thus, a fan can be tested in different designs of rig, all of which satisfy the criteria specified in the appropriate standard. An obvious question is what influences do the differences in installation have on the measured sound levels?

This question has been addressed as one aspect of a series of tests which were undertaken at six different laboratories within Europe as part of a programme supported by the Community Bureau of Reference of the EEC. The research programme was aimed at experimentally evaluating the accuracy with which fan in-duct sound power levels could be determined and

included an assessment of effects of installation variations on the measured levels.

The participating laboratories were CETIAT in France, DFVLR in Berlin, FRG, the Universite Catholique de Louvain (UCL) and the Universite Libre de Bruxelles (ULB) in Belgium and Woods of Colchester Ltd and NEL in the UK. The project was co-ordinated by the National Engineering Laboratory (NEL).

This paper describes the experimental project and presents information relating to the evaluation of any significant installation effect on sound power levels.

## 2 OUTLINE OF THE EXPERIMENTAL PROGRAMME

The programme was based on the experimental determination of the in-duct sound power levels of two fans from measurements of sound pressure level undertaken according to the procedures specified in the Draft International Standard DIS 5136(4). Two fans, one axial and one centrifugal both typical of those in use for general ventilation applications, were tested in each participant's laboratory.

There were two sets of tests. Firstly, each of the participating laboratories tested the two fans in a common duct installation, known as the 'Reference Rig'. This reference rig, which was designed and manufactured by the National Engineering Laboratory specifically for the intercomparison test programme, was shipped to each laboratory along with the fans to ensure that they would be tested by each laboratory under identical installation conditions. The information from this set of tests would reveal the level of inter-laboratory repeatability.

Secondly, each laboratory tested both fans in rigs of their own construction designed to comply with the specifications of the Draft Standard. This was done to evaluate whether any additional uncertainty was introduced into the measured sound power levels by variations in rig geometry. Tests were carried out at NEL at the start and finish of the project.

## 3    DESCRIPTION OF TEST FANS

The axial flow fan and the centrifugal fan were commercially available ones and were chosen as being representative of fans commonly used for heating, ventilating and light industrial purposes. Both fans had 600 mm diameter inlet and outlet apertures.

A sketch of the axial fan is shown in Figure 1a. The fan was manufactured by Airscrew Howden Ltd, type No 8BP600/22. It was fitted with eight blades in unequally spaced pairs with blade angles set to 22 degrees. The fan had no inlet guide vanes, stator vanes or diffuser. The motor, which drove the fan at 2900 r/min, was mounted in the airstream supported on a flat platform mounted off-centre across the airway.

Figure 1b is a sketch of the centrifugal fan. This unit was manufactured by Davidson & Co Ltd, type No A62B. The backward-curved impeller was driven via a vee-belt drive from a plinth-mounted motor. Both motor and drive were out of the airstream. The centrifugal fan was defined as including the 25 mm inlet interface flange and the outlet square-to-round transition piece as indicated in the sketch.

## 4    DESCRIPTION OF THE TEST INSTALLATIONS

### 4.1    General

The Draft International Standard specifies limits on the permitted diameter of the circular measurement ducts, the lengths of the measurement ducts and the sizes of any transition ducts which may be required to attach the measurement ducts to the fan under test. The acoustic performance of the inlet and outlet anechoic terminators is specified, but not their design or method of construction. Neither is there any particular design of throttle though it is specified that it must be situated on the far side of the anechoic termination from the measurement duct and that the level of any noise generated by the flow through the throttle is to be at least 10 dB below the level of the fan generated sound pressure levels in the measurement duct. No specific method is specified for the determination of fan flowrate. There is, therefore, potential for having a considerable range of test rig constructions for undertaking in-duct noise measurement on any one fan.

### 4.2    The Reference Test Rig

Figure 2 shows the arrangements of the reference rig for the tests conducted on the axial fan.

Details of the test rigs are given below.

The diameter of the measurement ducts was selected to be identical to the fan inlet and outlet diameters which were all 600 mm. No conical transitions were needed and only an intermediate duct for the outlet side was included in the arrangement. The ducts were manufactured of 3 mm thick mild steel. Access

hatches to give a means of installing the in-duct microphones were provided at appropriate locations in the inlet and outlet ducts. An inlet cone was manufactured to the requirements of BS 848 : Part 1 : 1980 in order to provide the means of determining the airflow. The mean internal diameter of the inlet cone was 599.3 mm.

The inlet and outlet anechoic terminations were designed using the technique outlined in Reference 4 to achieve a performance significantly better than the limiting values specified in the Draft Standard. A catenoidal profile was chosen for the expansion and this was approximated by a series of conical sections. The rate of flare on the inlet side of the inlet termination was twice that of the expansions on the fan sides of both inlet and outlet terminations. The acoustic lining material was an expanded polyurethane foam with the trade name 'Barafoam' which is manufactured by Kay Metzeler Ltd. The foam density was approximately 32 kg/m$^3$.

The disc throttle, mounted at the extreme end of the outlet termination was lined with Barafoam material formed into a truncated cone section. In order to control the fan flowrate, the axial location of the disc could be altered by an electric motor driving a threaded shaft.

The acoustic performance of the outlet terminator on the reference rig is shown in Fig. 3 as the trace labelled BCR. Both of the reference rig terminations were well within the performance specified in the Draft Standard.

### 4.3    Laboratory Test Rigs

The diameters of the measurement ducts used in the rigs constructed by the individual laboratories are tabulated below. Sketches showing the inlet and outlet connections to the fans are presented in Fig. 4.

Diameters of Measurement Ducts
in Laboratory Rigs

| Laboratory | Measurement Duct diameter (mm) |
|---|---|
| DFVLR | 500 |
| ULB | 550 |
| CETIAT | 630 |
| NEL | 710 |
| Woods | 800 |
| UCL | 800 |

The smallest test duct diameter, 500 mm, is the minimum permitted for a 600 mm diameter fan. The maximum allowable measurement duct diameter is 848 mm.

### 4.4    Performance of Anechoic Terminations

The acoustic performance of all of the anechoic terminations was evaluated using a procedure specified in the Draft Standard.

Fig. 3 presents the measured reflection coefficients of the outlet side anechoic terminations used in the test rigs assembled by the individual laboratories along with the maximum values of pressure reflection coefficient allowed by DIS 5136.2. All of the terminations met the specified limits.

## 4.5 Checks on signal-to-noise Ratios and Flow Conditions in the Ducts

The Draft International Standard stipulates that, as a preliminary check, the signal-to-noise ratio of fan sound to turbulence noise in the test duct should be measured. This was done by measuring in-duct sound pressure levels firstly with the microphone fitted with a nose cone then with the turbulence screen. The level difference between the sound pressure levels measured with the two windscreens indicated that there was at least a 6 dB signal-to-noise ratio between the fan generated sound and the turbulence induced noise even in the smallest diameter test ducts, the 500 mm diameter ducts of DFVLR.

As part of the signal-to-noise ratio check, the airflow velocity at the microphone position was measured. This was done at NEL by inserting a pitot tube into the duct at the position of the microphone. In addition to the airflow velocity, this measurement also yielded an estimate of the swirl present in the airflow. The swirl angle of the flow was found to be 10° or less in the inlet duct for all test rig configurations and in the outlet duct for all test configurations of the centrifugal fan. In the case of the axial fan in the outlet duct, however, substantial swirl velocity components were present with swirl angles as large as 45° being measured at the lowest test flowrates. The signal-to-noise ratio was above the minimum specified value of 6 dB throughout the tests.

## 5 AVERAGED SOUND POWER LEVEL

In the test series at each laboratory, both fans were operated at six specified flowrates. At each flowrate sound pressure levels were measured at three circumferential positions in the test ducts in accordance with the procedures specified in the standard. Additional tests were undertaken to investigate other factors including the repeatability of noise measurement within each laboratory, the calibration of microphones and measurement systems and the variations between individual turbulence screens.

Within each test laboratory the repeatability of three or four sets of sound pressure measurements taken over a short period was generally such that sound power levels were within about 0.5 dB of the mean. When the time between repeats was extended to several days the spread of results increased but was normally less than 1.0 dB.

In order to determine the degree of repeatability amongst the measurements performed by all the participating laboratories, averages and standard deviations were calculated of the sound power levels for all flowrates and all frequencies. The sound power levels reported by each of the laboratories were arithmetically averaged to produce averaged inlet and outlet spectra at each test flowrate for the following:

The axial fan in the reference rig,

the centrifugal fan in the reference rig,

the axial fan in the laboratories' own rigs, and

the centrifugal fan in the laboratories' own rigs.

Figs 5 and 6 show the average levels for the best efficiency flowrates of the axial fan (5.1 m$^3$/s) and the centrifugal fan (3.5 m$^3$/s) respectively.

It is clear that for both fans there are significant differences between the inlet and outlet sound power spectra. For the centrifugal fan the outlet sound power is approximately 4 dB greater than the inlet at all frequencies up to 2 kHz. In the case of the axial fan the outlet sound power is apparently almost 10 dB higher at low frequencies up to 160 Hz but above 500 Hz the inlet noise is greater, being about 4 dB higher at 500 Hz and 10 dB higher at 10 kHz.

The average sound power levels measured for the centrifugal fan in both the reference and laboratory rigs are fairly similar and, as will be shown later, the differences are well within the expected uncertainties. There are larger differences between the two sets of sound power data from the axial rigs though, again, these are within the limits of uncertainty derived from the test work.

## 6 STANDARD DEVIATIONS OF SOUND POWER LEVELS

### 6.1 Calculation of Standard Deviations

Once the average sound power levels had been calculated, each of the original data sets from each laboratory was compared with the averaged value to determine the size of the differences, frequency band by frequency band for each test flowrate.

The standard deviations of the average sound power levels were then calculated for inlet and outlet noise of each fan installation across the basic data sets for each individual third octave frequency and at each individual test flowrate. Each of these individual standard deviations was therefore based on seven measured test values since NEL had tested the fans on two occasions. For example, the seven individual values of sound power level at 100 Hz in the inlet duct from each laboratory for the centrifugal fan operating at a flowrate of 3.0 m$^3$/s were averaged to yield a mean sound power level and its related standard deviation for that fan at that particular flowrate in that measurement duct at that frequency.

Comparison of the standard deviation spectra for each installation configuration showed that the deviations were essentially independent of flowrate. It was therefore appropriate to derive overall standard deviations by integrating across all test flowrates to yield values which could be compared with the expected measurement standard deviations given in Table 1 of DIS 5136.2 for tests conducted to its specifications.

These overall measurement standard deviations are presented in Figs 7 and 8 for the axial and centrifugal fans respectively, along with the expected standard deviations quoted in DIS 5136.2.

The repeatability of the measurement is poor at high frequencies for both fans. It is believed that some of this is due to the high attenuation of sound pressure by the microphone turbulence windscreens causing the frequency analysers to operate close to the noise floor of their dynamic ranges where the resolutions of the instruments are becoming poor. This belief is supported by the observation that the deterioration in accuracy begins at a lower frequency in the case of the centrifugal fan than the axial fan, occurring at 4 kHz rather than at 6.3 kHz. This may be partly due to differences in the shape of the noise spectra of the two fans. The high frequency end of the centrifugal fan noise spectra fall to low levels quite steeply which would lead to less accurate estimates of the levels being made by frequency analysers operating near the floor of their dynamic range.

## 6.2  Variations between Inlet and Outlet Standard Deviations

In the case of the centrifugal fan the average measurement standard deviations are much the same irrespective of whether the measurements are made in the inlet side duct or in the outlet side duct. In the case of the axial fan, however, the average measurement standard deviations at the lower end of the frequency range are very much greater in the outlet duct than in the inlet duct.

The most probable cause of this difference is the presence of large amounts of swirl in the duct downstream from the axial fan. High swirl components in the in-duct airflow are known to adversely affect the performance of slit-tube turbulence screens and this may be a major factor responsible for the increased deviations observed in this case. The degree of swirl present in the flow downstream of the axial fan is sufficiently large as to be outside the limits set in the Draft International Standard, even although the measured signal-to-noise ratio was satisfactory.

## 6.3  Comparison between Reference Rig and Laboratory Rigs

A feature of the data shown in Figures 7 and 8 is the similarity between the deviations obtained for the tests conducted in the reference rig and those derived from measurements made in the various laboratories' own rigs. The average standard deviations of the tests in the laboratories' own rigs might have been expected to be greater due to the effects of variations in the geometry of the rigs and the performances of their anechoic terminations. Also, elimination of the transition duct from the axial fan discharge diameter to the test duct diameter should have eliminated one source of potential uncertainty (6) from the reference rig. No clear trend of this sort is observed. Overall, this suggests that the amount of additional uncertainty introduced into measured sound power levels by variations in rig construction within the limits allowed in the Draft International Standard is small compared to other sources of error.

## 6.4  General Comments

Over the majority of the frequency range the standard deviations for the centrifugal fan are within the values presented in the Draft Standard. The values were estimated by Bolleter et al (6) essentially from theoretical considerations. Excluding the data affected by the swirling flow in the outlet duct of the axial fan, the standard deviations derived from the measurements of the axial fan noise lie close to the estimated uncertainties.

Each value of standard deviation presented in Figures 7 and 8 has been derived from 42 separate values of sound power level, each of which was calculated from three independent measurements of sound pressure. The standard deviations are thus good indications of the scatter likely to be encountered whenever the procedures given in the Standard are followed.

It should be noted that the standard deviations given in the Draft Standard and the standard deviations derived from the test series reported in this paper are the standard deviations of many independent sound power evaluations made in compliance with the standard and are not the likely error associated with any single assessment of sound power. If the sound power level is based on one test the value might be 2, 3 or even more standard deviations away from the 'true' value. During the analysis of the test data there were many instances where individual sound power levels were more than 5 dB away from the mean values.

## 7  EFFECT OF TEST DUCT DIAMETER

The absence of any clear trend towards higher overall measurement deviations in the data set obtained from the laboratories' individual rigs when compared to the data set obtained from tests in the reference rig alone, suggests that rig size and shape variations within the DIS 5136.2 limits have only minor effects on the measured levels. However, it appeared from an initial brief analysis that the levels measured in ducts of smaller diameters seemed to be slightly higher than those measured in the larger ducts.

In order to test this, the differences between the sound power levels measured in each of the laboratories' own rigs and the average of the values obtained in the reference ducting were calculated. These differences were calculated, point by point, for each frequency and flowrate for both inlet and outlet measurements. These differences were then averaged to give a single mean difference for each measurement duct diameter. These mean differences are plotted as a function of duct diameter in Figures 9 and 10 for the axial and centrifugal fans respectively.

The trends indicated on these figures are not particularly well defined, especially when the size of the known uncertainties of measurement are considered. However, a general reduction in apparent sound power level with increasing measurement duct diameter of approximately 2 dB over the range of diameters tested, is suspected.

## 8 CONCLUSIONS

The scatter of measurements within each of the individual laboratories was generally within $\pm 1$ dB and often within $\pm 0.5$ dB.

Individual differences from the mean value in excess of 5 dB have been observed both in the laboratories' rigs and in the reference rig. The overall standard deviations for the measurements made in the reference rig were found to be comparable to those obtained from the laboratories' own rigs and in some cases worse. It is concluded that errors introduced by variations in the geometry of the test rigs and the performance of their anechoic terminations within the limits allowed in the Draft International Standard are small compared to other sources of uncertainty.

In the case of the centrifugal fan, most of the measurement standard deviations were within the limits quoted in the Draft International Standard. The main exceptions to this occurred at high frequencies where low signal levels were approaching the noise floor of the measurement instrumentation.

In the case of the axial fan a majority of the measurement standard deviations lie outside the limits quoted in the Draft Standard. This was most pronounced for the outlet side measurements at lower frequencies. It is considered likely that high levels of swirl in the outlet duct are responsible for these increased measurement uncertainties.

The installation of an intermediate duct of the same diameter as the fan inlet between the fan and an upstream transition section improves the repeatability of the sound power measurement.

The findings from this series of tests will be used to assist in the preparation of the ISO Standard for fan noise measurement.

## ACKNOWLEDGEMENTS

The authors wish to acknowledge the help given by colleagues at all the participating test laboratories.

The tests at all the laboratories were partly funded by the Bureau Community of Reference of the EEC.

This paper is contributed with the permission of the Director of the National Engineering Laboratory, Department of Trade and Industry. It is Crown copyright.

## REFERENCES

(1) BAADE, P. Effects of acoustic loading on axial flow fan noise generation. Noise Control Engineering, January-February 1977, 8(1), 5-15.

(2) CREMER, L. The treatment of fans as black boxes. Second Annual Fairy Lecture. J. Sound & Vibr., 1971, 16(1), 1-15.

(3) BOLTON, A. N. and MARGETTS, E. J. The influence of impedance on fan sound power. Paper C124/84, Installation Effects in Ducted Fan System, I.Mech.E., London, May 1984.

(4) SECRETARIAT OF ISO TC 43/SC1/WG3. Acoustics - Determination of sound power radiated into a duct by fans - in-duct method. ISO DIS 5136.2, 1984. DIN Burggrafen Strasse 4.10 D1000 Berlin 30.

(5) BOLTON, A. N. and MARKETTS, E. J. Anechoic terminations for in-duct fan noise measurement, Paper H1. Int. Conf. on Fan Design and Applications, Guildford, England, 7-9 September, 1982. BHRA, Cranfield, Bedford.

(6) BOLLETER, U., COHEN, R. and WANG, J. Design considerations for an in-duct soundpower measuring system. J. Sound Vib., 1973, 28(4), 669-685.

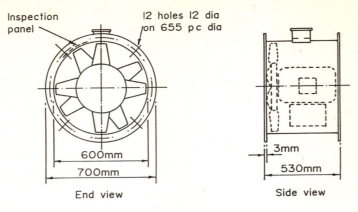

(a) Sketch of reference axial fan

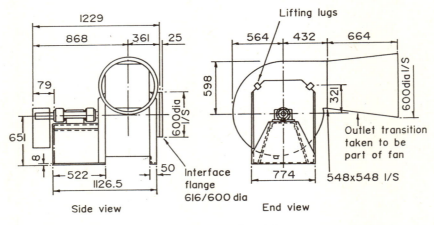

(b) Sketch of reference centrifugal fan

Fig 1     Sketches of the two test fans

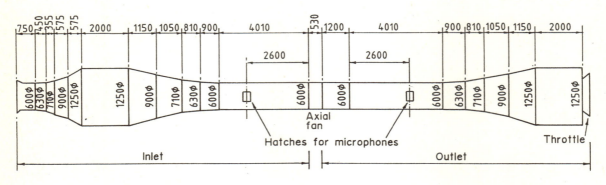

All dimensions in mm

Fig 2     BCR reference test rig for the axial flow fan

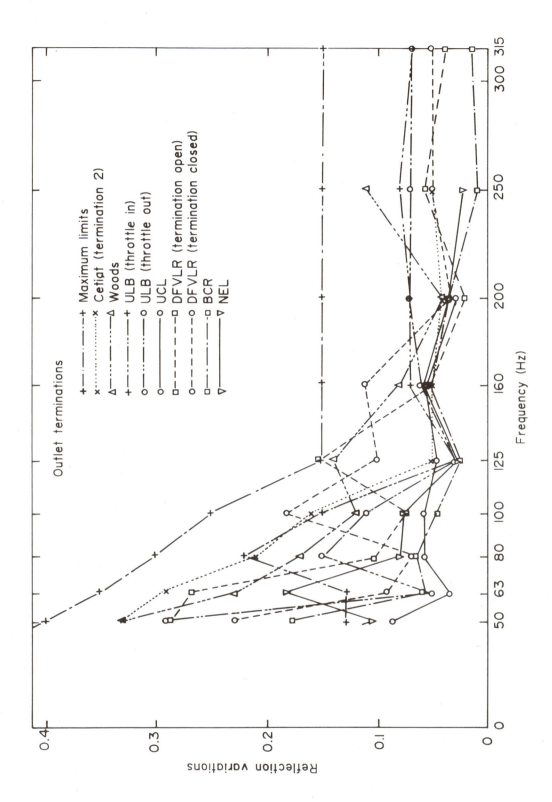

Outlet terminations

- +  Maximum limits
- ✕  Cetiat (termination 2)
- △  Woods
- +  ULB (throttle in)
- ○  ULB (throttle out)
- ○  UCL
- □  DFVLR (termination open)
- ○  DFVLR (termination closed)
- □  BCR
- ▽  NEL

Fig 3    Measured performance of the laboratory rig outlet anechoic terminations

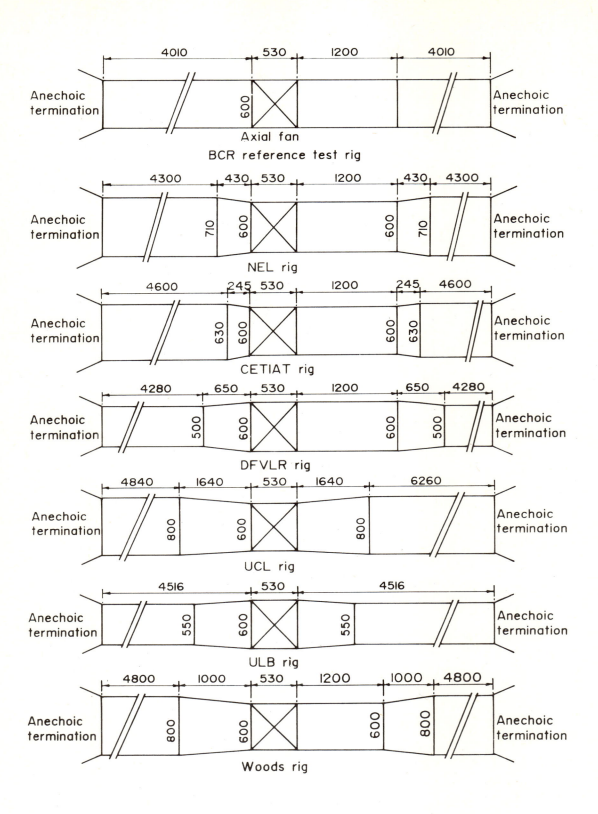

Fig 4    Dimensions of laboratory test rigs

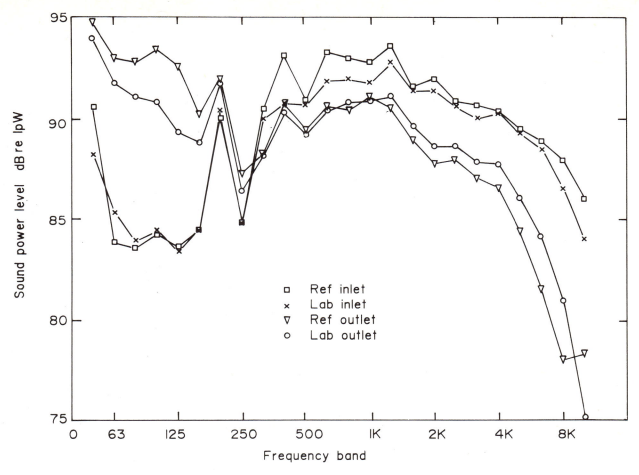

Fig 5    Axial fan — mean sound power levels

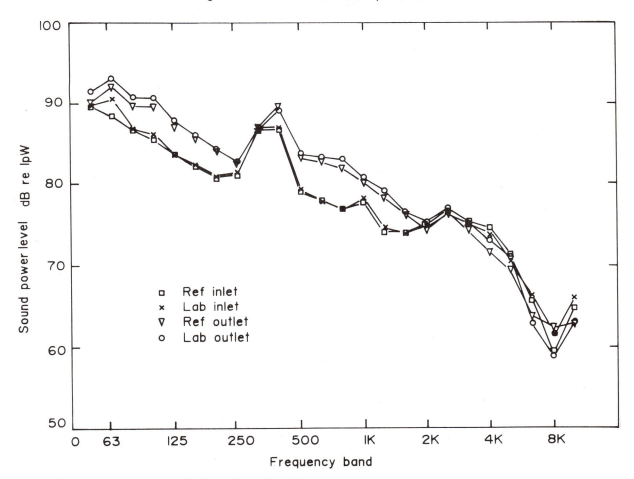

Fig 6    Centrifugal fan — mean sound power levels

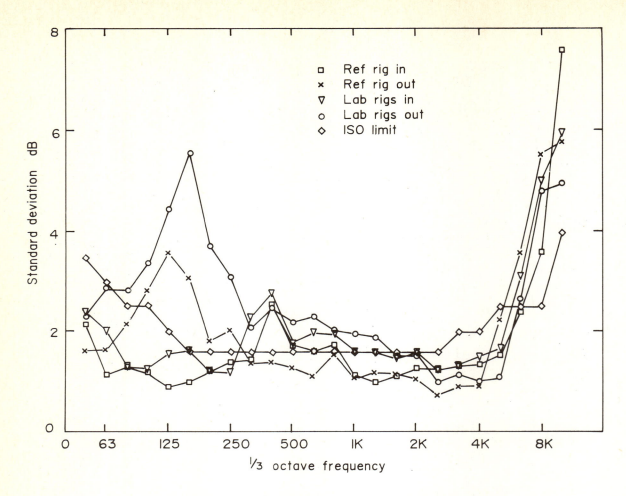

Fig 7    Overall standard deviations for axial fan

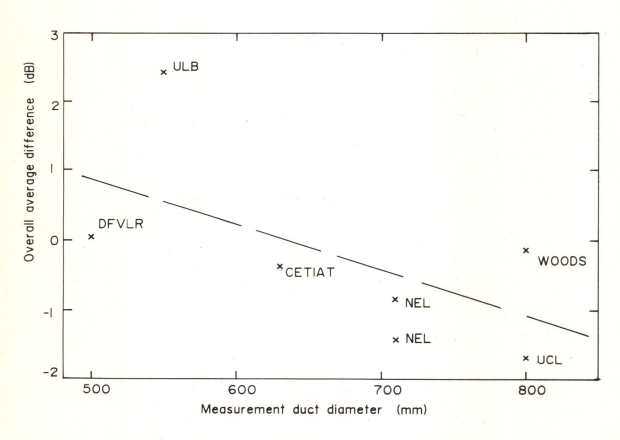

Fig 9    Differences between sound power levels measured in laboratory
rigs and mean sound power levels in reference rig: axial fan

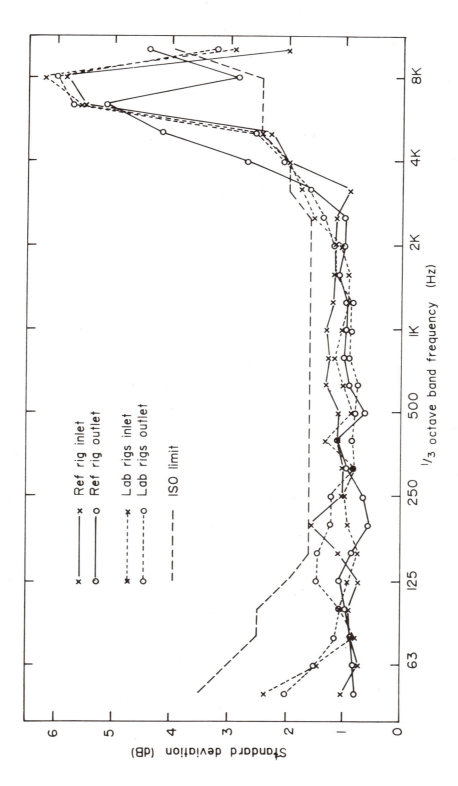

Fig 8    Overall standard deviations for centrifugal fan

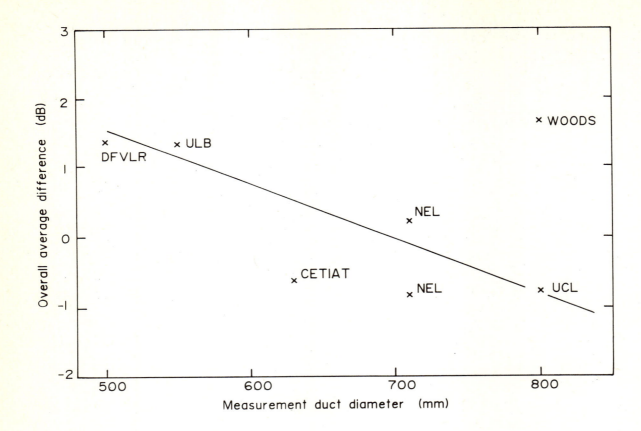

Fig 10    Differences between sound power levels measured in laboratory
          rigs and mean sound power levels in reference rig: centrifugal fan

# Testing a range of fans to BS 848: Part 2: 1985—a manufacturer's viewpoint on the resulting sound information and its use within published data sheets

C W LACK, BSc
Elta Fans Limited, Byfleet, Surrey

SUMMARY    This paper outlines the testing carried out on a range of non geometric adjustable pitch axial flow fans, so as to establish their acoustic performance.  The testing was carried out to the latest requirements of BS 848 Part 2(1985) - reference 1, which is the current fan sound testing standard.

Upon completion of the onerous test programme the "acoustic footprint" of the fan range was derived.  Then the most accurate and unambiguous means of presenting this data was evolved, in line with the recommendations for fan performance publication as in reference 2.

The present requirement for fan performance certification within the technical schedule of BS 5750 means that the publication of sound data of traceable and certifiable origin will take on greater importance.  With this requirement now upon us, sound data of considerably fuller and more meaningful form will become the norm.

This fan range is the first one to have been tested and to have had its sound data fully illustrated since BS 848 Part 2(1985) has been published.

Within the test programme which was predominately carried out by an independent accredited laboratory, a number of special investigations were carried out; some of which are highlighted within this paper.

## 1   INTRODUCTION

The testing of fans for acoustic performance has taken a major advance in both accuracy and complexity following the publication of BS 848 Part 2(1985).  Prior to that testing had been carried out to the earlier (1966) revision, which besides advocating measurement in octave band width, resulted in lower accuracy in the 63Hz band.

With the greater understanding now of fan sound propagation and the acceptance that installation differences adjacent to the fan can cause differing resultant sound levels; (ie fan impedance, see reference 4) the latest revision to Part 2 was drawn up.

This latest standard further clearly states that when testing a range of fans whether of geometric or non-geometric family, then adequate tests should be carried out to establish the laws pertaining to the fan range with respect to both size and speed variation.  The degree of inter-polation or extrapolation is limited so that good accuracy is established in the "sound footprint" of the family of fans under test. This of course being the ultimate aim.

The 1985 version of the standard results in third octave band sound levels being available with the 63Hz level attaining the same order of accuracy as other bands.

Axial fans, when produced in the most efficient and economic fashion, generally have adjustable aerofoil blades mounted on a range of differing hub diameters, thus achieving the necessary hub/tip ratios.  By further varying the blade numbers within these hubs the necessary range of solidities is achieved.

Sometimes within an overall range there are size groupings utilising the same major components with varying diameters being achieved by tip cropping.  In the fan series illustrated in this paper, even greater utilisation of impeller components is achieved by the use of one blade profile within the overall size range.  Obviously at the design stage careful consideration was undertaken to ensure that this optimisation of "hardware" did not compromise too greatly the aerodynamic or acoustic properties of the overall range.

## 2   TEST CONSIDERATIONS

Within the current BS 848 Part 2 standard various test rig layouts are permitted, which result in either In-duct or Reverberant/ Semi-reverberant room test approaches being adopted.

The decision as to which test approach to follow depends greatly on the type of product to be tested and the range of products likely to be tested within the facility.

In-duct testing was followed in the test series on these axial fans. This allowed the simultaneous measurement of both the inlet and outlet side sound levels. Figure 1 shows the test arrangement for the 710 mm series of fans. The length of this rig was approximately 20 metres, when testing the higher rotational speed units. At lower speeds a booster fan section complete with its own attenuator of some 8.5 metres length was further required, to enable measure - ments to be achieved close to the free-air delivery condition.

Prior to 1985 when the aerodynamic testing was underway on this family of fans, (reference 5), it was not clear as to which revision of BS 848 Part 2 that the range would subsequently be tested. It was however quite clear that the simultaneous testing of aerodynamic and accoustic properties of the range was not permitted. However it was still thought to be a worthwhile exercise to take sound measurements in-duct on the outlet side of the fans while they underwent their aerodynamic testing. The microphone being located generally to the requirements of the 1966 standard. Whilst it was realised that end reflection effects could bear heavily on the accuracy of the overall levels, especially on the smaller diameters; it was hoped that the "spectrum profiles" so obtained would prove useful in the subsequent sound analysis.

Figure 2 shows the microphone position during these tests, together with typical results when compared to the subsequent 1985 revision results.

## 3  GENERAL TEST PROGRAMME

The test programme was set so that the number of fan variants tested was in accordance with the requirements of the British Standard. This required five different fan sizes to be tested namely 315,500,630,710 and 900 mm diameters, together with the manufacture and calibration of five complete duct arrange-ments.

The main aim of the test series was to amass sufficient test data so enabling the accurate prediction of intermediate size and speed units; within the overall family of fans. To enable this "acoustic footprint" to be obtained the empirical method as outlined in Appendix G of BS 848 Part 2(1985) was utilised. In this method the sound power levels for a given fan are calculated using the following equation.

$$L = Lg + 10.(6+a).\log n + 10.(8+2a+b).\log Dr$$

$$....(i)$$

In which n is the shaft speed, Dr is the impeller diameter and a,b and Lg are empirical scaling parameters. These being derived from the test results. The "a" coefficient being a scaling index associated with geo-metric scaling of the fan sound power with size and speed. The "b" coefficient is a further scaling index which allows for deviation from the true geometric scaling condition (as is the case with the fan family under consideration here). Lg is the frequency dependent term which represents a generalised spectrum shape for the fan series. L being the resultant sound power level.

## 4  GENERAL DATA BASE

To generate the required data base both inlet and outlet side in-duct one third octave band levels were taken on fan sizes as stated in section 3. Each size was run at three different rotational speeds with the impeller set in all of its basic con-figurations relative to that particular size.

Typical sound power measurements as achieved on a 900 mm diameter 14 bladed fan unit when set at a mid blade angle and running at 1440 rpm are shown in figure 3, together with the subsequently derived values using equation (i). These are plotted for various flows on a base of frequency/ rotational speed.

The airflow at which the impeller blade stalls is clearly shown on this figure as it helps to highlight the considerable change in spectrum profile that occurs when the blades are aerodynamically heavily loaded.

Having established the data banks covering both inlet and outlet side sound levels for all the hub/blade combinations, further analysis could be carried out to establish the laws relating to the variation of the overall sound power levels. Well known empirical relationships for changes in size and speed have been used in the past for carrying out sound adjustments. These being -

$$\Delta\, dB = 55\, \log n_1/n_2 \text{ (for speed)}$$
$$...(ii)$$
$$\text{or } \Delta\, dB = 55\, \log D_1/D_2 \text{ (for size)}$$

Using the data bank sound information, overall sound power levels were derived; an example of which is plotted in figure 4. These figures show typical curves for the variation in Lw with both size and speed. Such results were used in the derivation of the coefficient 'k' for use in the relation-ship

$$\Delta\, dB = k.\, \log X_1/X_2 \qquad ...(iii)$$

Where $X_1$ and $X_2$ are the initial and final values of either size or speed.

The following general coefficients were found to be applicable.

| Hub Code | 'k' | |
|---|---|---|
| Hub (mm)/Blades | Size | Speed |
| 150/5 | 55 | 40 |
| 150/10 | 55 | 48 |
| 250/7 | 50 | 42 |
| 250/14 | 58 | 48 |

Analysis of much of the published fan performance data to date will show that the empirical coefficients as shown in equation (ii) have been widely used. One should now expect these different coefficients to more frequently appear. Indeed it may well be the case, that coefficients of considerably smaller value for some types of fans to those indicated above, will occur.

## 5 EFFECT OF TIP CLEARANCE VARIATION

The variation of airside performance with varying tip clearance had been derived during the aerodynamic testing of the fan range. A similar series of tests were carried out during the sound tests so as to establish their variation under similar circumstances. Figure 5 shows the variation for both the airside and the average sound levels across the spectrum, as tip clearances were increased from the stated datum value when operating within the designated performance area. Naturally the minimum In-duct sound levels were observed at minimum tip clearance. Towards the free discharge end of the characteristic minimal increase in sound levels were observed. However towards the high pressure end of the characteristic considerable increase was found.

The 315 and 400Hz band widths were found to be particularly affected, with increase of as much as 17 dB being recorded under the maximum tip clearance conditions tested.

These tests highlighted the importance of good tip clearance control on impellers designed to run in the ducted mode. Such impellers are heavily loaded in their tip regions, with resulting flow propagation across the blade tip, which results in severe noise when tip clearances become too great.

## 6 EFFECT OF FAN FORMS A & B

During the test programme a series of comparative tests were carried out to determine the effect of fan form on the measured sound levels, both upstream and downstream of the installed fan. The normal arrangement for long cased fan units is that where the airflow passes through the impeller before passing over the drive motor (ie form B). The alternative form A, is when the

motor is placed on the upstream side of the impeller. In such an arrangement the airflow can be affected by the presence of the motor platform or supports when foot mounted electric motors are utilised.

Figure 6 shows comparative one-third octave band levels obtained on a half solidity fan unit of 900 mm diameter when running at 1440 rpm. It can be seen that especially on the inlet side considerable variation in the resultant differences occur for the two operating points shown.

The general outcome of the investigation was that for octave bands up to about 500Hz that an increase of 3 to 4 dB was applicable for form A configuration. Above 500 Hz the difference in most cases was of a minor order. The resultant overall variation between fan forms was of the order of 2 to 3 dB.

## 7 EFFECT OF IMPELLER BLADE MATERIAL

Impeller blades are manufactured in two materials, die cast aluminium and injection moulded glass fibre reinforced polypropylene. Both these blade forms are of identical aerofoil section, hence when tested for air performance gave identical results.

Early in the series of sound test it was necessary to establish whether there was any real difference in their sound performance. If no differences were found then only planned comparative tests would be necessary, whereas if there was a real difference then complete testing of both materials over the entire test programme would be necessary.

On the 710mm series of tests full comparative tests were conducted on all impeller solidities and speeds. It was established that negligible differences occured. This is contrary to what one might have expected bearing in mind the differing resonant frequencies of the types of blade. The table below shows typical comparative figures for both the inlet and outlet in dB achieved on identical 10 bladed impellers of 710mm diameter at 1440 rpm.

Difference: Aluminium - Plastic

| Frequency Hz | Inlet side | Outlet side | Mean |
|---|---|---|---|
| 12.5 | .06 | .07 | .06 |
| 16.0 | - .02 | - .10 | - .06 |
| 20.0 | .17 | .06 | .11 |
| 25.0 | - .26 | - .22 | - .24 |
| 31.5 | - .21 | - .01 | - .11 |
| 40.0 | .07 | .05 | .06 |
| 50.0 | .02 | - .13 | - .05 |
| 63.0 | - .06 | - .11 | - .08 |
| 80.0 | .84 | .52 | .68 |
| 100.0 | - .25 | - .32 | - .28 |
| 125.0 | - .27 | - .36 | - .32 |
| 160.0 | .14 | .15 | .15 |
| 200.0 | - .51 | - .16 | - .33 |
| 250.0 | - .62 | - .33 | - .48 |

/....

....../ Cont'd

| Frequency Hz | Inlet side | Outlet side | Mean |
|---|---|---|---|
| 315.0 | - .99 | -1.26 | -1.13 |
| 400.0 | -1.61 | -1.46 | -1.54 |
| 500.0 | -1.56 | -1.51 | -1.53 |
| 630.0 | - .81 | - .79 | - .80 |
| 800.0 | - .31 | - .43 | - .37 |
| 1000.0 | - .53 | - .34 | - .44 |
| 1250.0 | - .89 | - .91 | - .90 |
| 1600.0 | - .76 | - .90 | - .83 |
| 2000.0 | - .82 | - .81 | - .81 |
| 2500.0 | -1.38 | -1.09 | -1.24 |
| 3150.0 | -1.52 | -1.39 | -1.45 |
| 4000.0 | -1.57 | -1.41 | -1.49 |
| 5000.0 | -1.44 | -1.20 | -1.32 |
| 6300.0 | - .48 | - .63 | - .56 |
| 8000.0 | - .10 | - .32 | - .21 |
| 10000.0 | - .12 | .69 | .28 |

Overall Mean    - .51 dB

## 8  VARIATION IN BLADE INCIDENCE

Another necessary series of tests was to
establish the variation of sound levels
with blade incidence or angle.  Hence a
number of impeller combinations had their
blade angles adjusted during the test programme.

Typical resultant sound levels are
shown in figure 7 for three differing blade
angles for a 710mm impeller having 10 blades
at 4 pole speed.

Such tests together with the wealth
of sound information gained during the airflow
tests, added to the data bank of sound information.  It quickly became evident that
the sound levels generated by the impellers
as their blade angles were changed did not
follow any simple rule.  But that they resulted
in a complex relationship of flow, blade
angle and aerodynamic loading.

This interactive sound pattern would
subsequently prove to make publication of
the sound data more complex than had been
anticipated.

## 9  PUBLICATION OF PERFORMANCE DATA

Having produced large banks of sound information the next step was to convert these
sound figures into a meaningful format for
fan performance curve presentation.  As
mentioned in the introduction, the recommendations of reference 2 gave guidance as to
the format that the performance curve should
take, especially so far as the airside performance was concerned.

On the subject of sound data it recognises
that when testing to the new standard
additional data and possibly a change in
approach will be needed.  However up to
that time no one had tested extensively
to the new revision, hence they had not
come across the dilemma associated with

trying to simplify the sound data to any
great extent.

One has also to bear in mind the eventual
certification requirement even for the sound
data, as mentioned in the summary.  Hence
the data stated on performance sheets had
to be correct, not greatly interpolated
or extrapolated and had to be traceable
back to the actual test measurements.

Those conversant with acoustic analysis
will realise that as soon as one starts
to average or to take mean values between
varying sound spectra then the resulting
spectrum soon looses all similarity to
the original spectrum profiles.  If one
followed that philosophy, then subsequent
proving as to the origin of the sound data
with any real conviction would be difficult.

Hence the sound data of certain fans
was studied critically to arrive at a
form of sound presentation that truely
reflected the achieved test measurements.
It soon became evident that within the performance envelope covering the normal operating
blade angle range that like the airside,
the sound data was in a  constantly changing
or "live" situation.

In the past whilst overall levels have
been shown as spot values or contours on
performance data sheets, generally only
one sound spectrum has been indicated.  This
has normally been the spectrum associated
with the fan operating at or near its best
efficiency point.  However for economic
reasons axial fans are usually operated
in higher pressure regions, ie closer to
the peak pressure or stall cut-off line.
Of course they are also sometimes operated
at the other extreme, ie close to the free
air delivery or maximum flow condition,
as with propeller fans.

The possibility of having three spectra
quoted for each performance envelope was
investigated, whether these be low, medium
or high blade angle; ie varying flow conditions.
Or possibly three zones based on low, medium
and high pressure regions.

It soon became evident that whilst
both of these approaches began to help,
that a more sophisticated approach needed
to be considered.  Hence the 'nine sound
zone' approach was born as shown in figure  8.
This allowed spot values of the overall
sound power levels to be shown on the performance envelope in the normal way - with interpolation if required.  Together with nine
spectra which are those applicable at the
centre point of each zone.  This way the
data as presented is accurate, without interpolation and even more important certifiable
in the future with respect to reference
3 considerations.

Both inlet and outlet sound levels
are then available to the user with equal
accuracy on this new curve format.  Having
taken this step, it will be interesting
to see how the next fan manufacturer displays
his results; following sound testing to the

© IMechE 1990 C401/026

latest Part 2 standard.

REFERENCES

1  BRITISH STANDARDS INSTITUTION.  Fans for
   general purposes.  Part 2: Methods of noise
   testing. BS 848: Part 2: 1985.

2  FAN MANUFACTURERS ASSOCIATION.  Fan
   Catalogue performance data, second edition
   December 1987.

3  BRITISH STANDARDS INSTITUTION.  Specification
   for design, manufacture and installation.
   BS 5750: Part 1: 1987.

4  A.N. BOLTON AND E.J. MARGETTS.  The
   influence of impedance on fan sound power,
   National Engineering Laboratory.  Conference
   paper C124/84 I.Mech.E

5  BRITISH STANDARDS INSTITUTION.  Fans for
   general purposes Part1 Methods of testing
   performance BS 848: Part 1: 1980.

SYMBOLS

| | |
|---|---|
| a | Sound index |
| b | Sound index |
| D | Diameter |
| dB | Decibel |
| dBW | Decibel sound power |
| Hz | Frequency (c.p.s) |
| k | Coefficient |
| kW | Power in Kilowatts |
| L | Sound level in dB |
| $L_g$ | Generalised sound level in dB |
| $L_W$ | Sound power level |
| n | Speed |
| Pa | Pressure in pascal |
| $P_dF$ | Fan dynamic pressure |
| PHI | Flow coefficient |
| $P_sF$ | Fan static pressure |
| $P_tF$ | Fan total pressure |
| $P_R$ | Impeller power |
| $q_v$ | Volume flow |
| t | Tip clearance |
| X | Designated value |

GREEK LETTERS

$\Delta$  Differences

SUBSCRIPTS

1  Initial value
2  Final value

ABBREVIATIONS

%  Percentage

ACKNOWLEDGEMENTS

   The author acknowledges the assistance
given by the National Engineering Laboratory
over certain photographic and artwork used in
this paper.

Fig 1    Sound rig for type D testing to BS 848: Part 2: 1985

(a) Test arrangement for 710 mm fan series

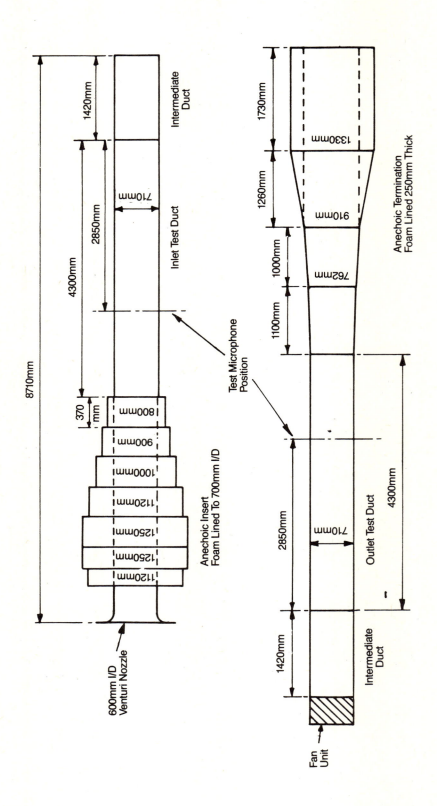

8710mm

1420mm — Intermediate Duct

710mm — Inlet Test Duct

2850mm

4300mm

370mm

800mm
900mm
1000mm
1120mm
1250mm
1250mm
1120mm

Anechoic Insert
Foam Lined To 700mm I/D

600mm I/D
Venturi Nozzle

Test Microphone
Position

1730mm

1330mm

1260mm

910mm

1000mm

762mm

1100mm

Anechoic Termination
Foam Lined 250mm Thick

710mm — Outlet Test Duct

2850mm

4300mm

1420mm — Intermediate Duct

Fan
Unit

(b) General dimensions of 710 mm fan test rig

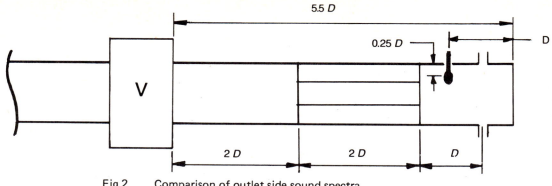

Fig 2    Comparison of outlet side sound spectra

(a) Location of microphone during type D aerodynamic tests

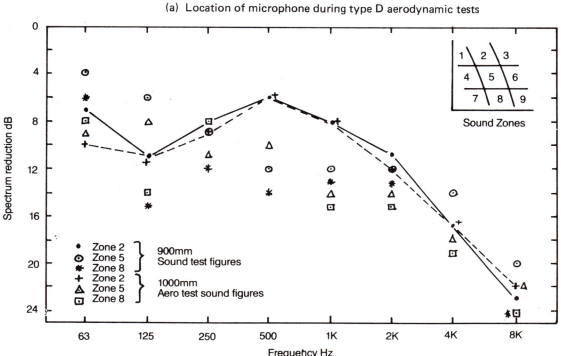

(b) Typical full solidity comparisons at 960 r/min

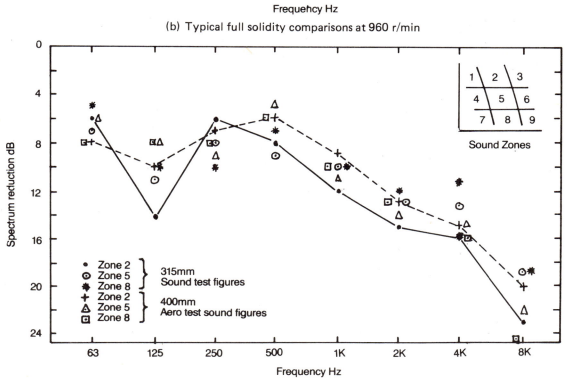

(c) Typical half solidity comparisons at 2880 r/min

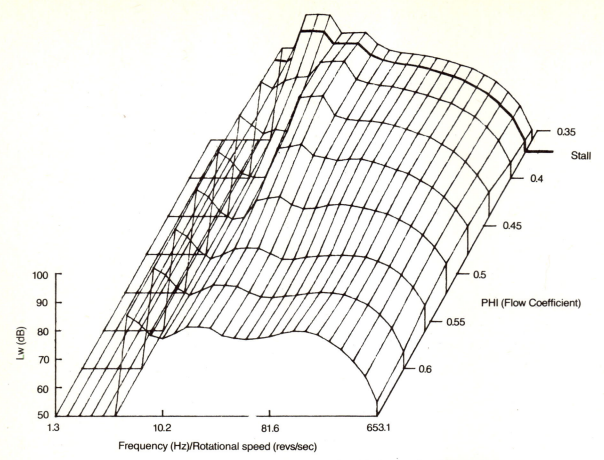

Fig 3 Comparison of measured and calculated induct outlet side sound
power levels for a 900 mm diameter 14 bladed fan unit at 1440 r/min

(a) Measured sound levels

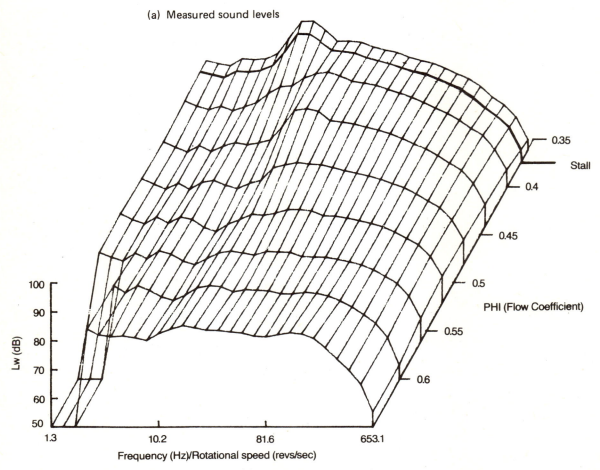

(b) Calculated sound levels

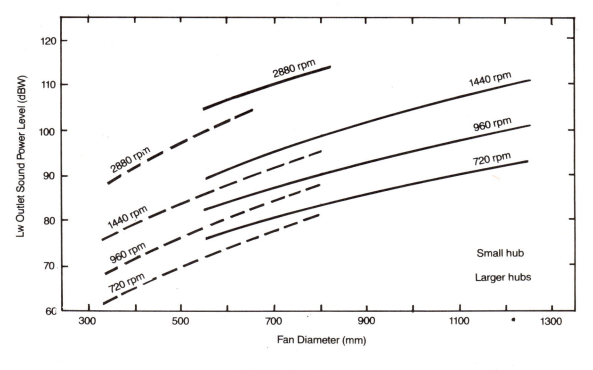

(a) Variation with size

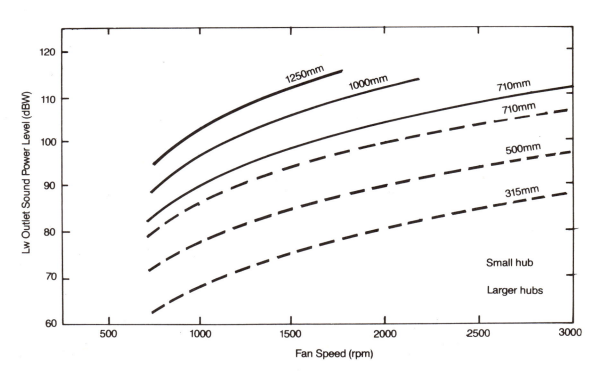

(b) Variation with speed

Fig 4    Overall induct sound power — full solidity fans

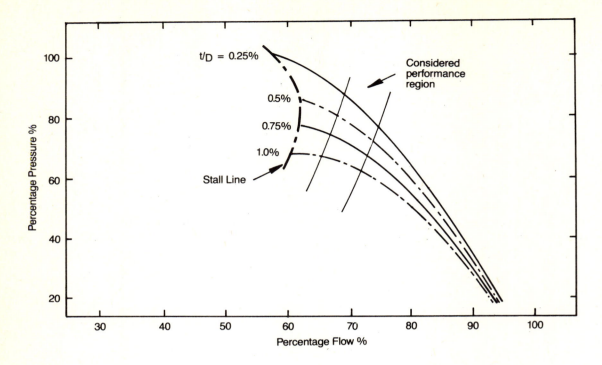

(a) Airside performance variations

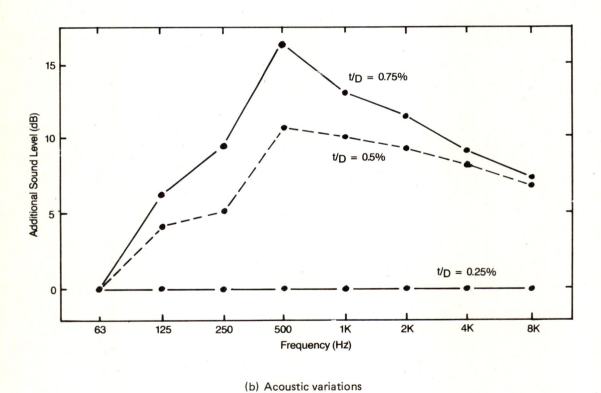

(b) Acoustic variations

Fig 5    Typical tip clearance variation — full solidity fan

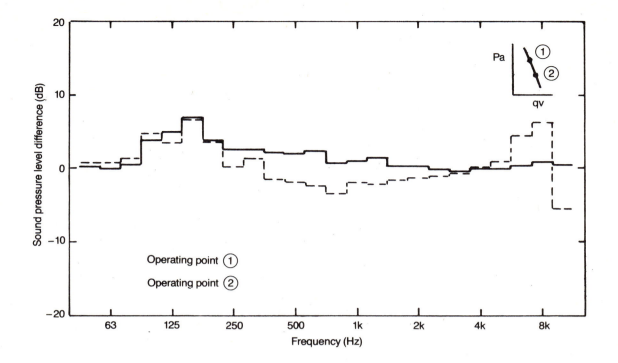

(a) Inlet side differences (form A—form B)

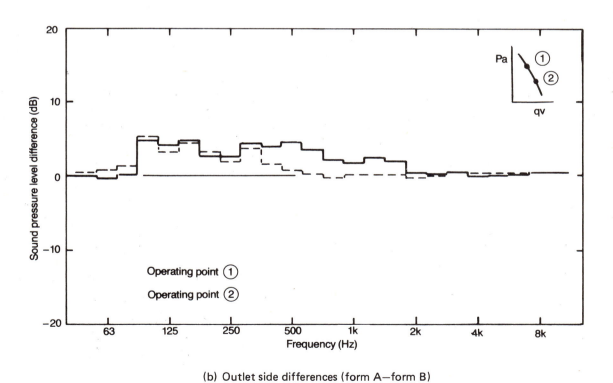

(b) Outlet side differences (form A—form B)

Fig 6    Comparative sound levels for a half solidity 900 mm at 1440 r/min

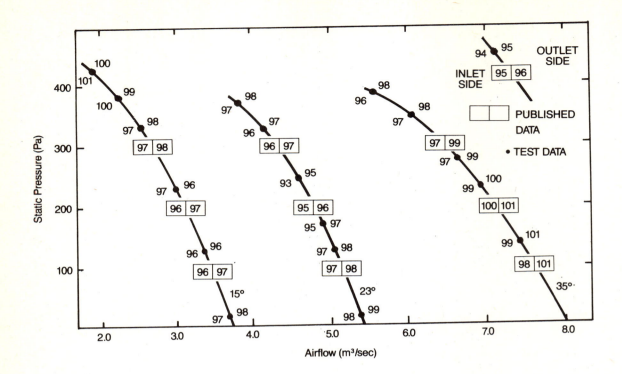

Fig 7    Variation of sound power levels with blade angle for a 710 mm full
         solidity fan at 1440 r/min

## Sound Data

**Axial Flow Fan**
**Type 630/150/10**
**1440 REV/MIN**

From the performance curve overleaf the Total Sound Power level In-duct on the outlet side of the fan is established for the duty condition, the fan. One can further see the Sound Zone (1 to 9) that the operating point falls with-in. By the use of the applicable zone table below other Total Sound Power levels may be found and then by applying the relevant spectra corrections the Octave band values may be deduced.

Sound Pressure Total and Octave levels may be obtained at a 1 metre distance (re. 20 μ Pa) assuming spherical radiation by the subtraction of 11 dB from the Sound Power figures. The total sound pressure level 'A' weighting may be found by applying the relevant 'dBA' correction shown in the tables below.

The correction values shown on this page are those applicable at the centre point of each zone. Sudden changes in total or similar octave band values do not occur across zone boundaries. Hence interpolation between adjacent zone figures can be made if required.

**Sound Zones**

| | | |
|---|---|---|
| 1 | 2 | 3 |
| 4 | 5 | 6 |
| 7 | 8 | 9 |

| Zone | Side | | Total | 63 | 125 | 250 | 500 | 1k | 2k | 4k | 8k | dBA |
|---|---|---|---|---|---|---|---|---|---|---|---|---|
| 1 | Inlet side | In-duct dB | +1 | -27 | -16 | -2 | -9 | -9 | -17 | -24 | -32 | -4 |
| | | Free Field dB | +1 | -35 | -30 | -17 | -17 | -9 | -17 | -24 | -32 | -4 |
| | Outlet side | In-duct dB | 0 | -16 | -14 | -8 | -8 | -8 | -13 | -18 | -29 | -4 |
| | | Free Field dB | 0 | -24 | -19 | -9 | -4 | -8 | -13 | -18 | -29 | -5 |
| 2 | Inlet side | In-duct dB | +2 | -17 | -20 | -9 | -4 | -6 | -10 | -14 | -26 | -2 |
| | | Free Field dB | +1 | -25 | -25 | -10 | -4 | -6 | -10 | -14 | -26 | -3 |
| | Outlet side | In-duct dB | 0 | -14 | -17 | -9 | -5 | -6 | -7 | -11 | -17 | -1 |
| | | Free Field dB | 0 | -22 | -22 | -10 | -5 | -6 | -7 | -11 | -17 | -1 |
| 3 | Inlet side | In-duct dB | 0 | -18 | -19 | -7 | -4 | -6 | -8 | -10 | -20 | 0 |
| | | Free Field dB | 0 | -26 | -24 | -8 | -4 | -6 | -8 | -10 | -20 | -1 |
| | Outlet side | In-duct dB | 0 | -7 | -11 | -8 | -9 | -9 | -9 | -13 | -20 | -5 |
| | | Free Field dB | -1 | -15 | -16 | -9 | -9 | -9 | -9 | -13 | -20 | -6 |
| 4 | Inlet side | In-duct dB | -2 | -12 | -17 | -3 | -6 | -8 | -8 | -19 | -24 | -4 |
| | | Free Field dB | -2 | -20 | -22 | -4 | -6 | -8 | -8 | -19 | -24 | -5 |
| | Outlet side | In-duct dB | 0 | -14 | -15 | -11 | -5 | -6 | -9 | -16 | -25 | -3 |
| | | Free Field dB | 0 | -22 | -20 | -12 | -5 | -6 | -9 | -16 | -25 | -4 |
| 5 | Inlet side | In-duct dB | -1 | -6 | -15 | -8 | -9 | -7 | -10 | -11 | -19 | -3 |
| | | Free Field dB | -1 | -14 | -20 | -9 | -9 | -7 | -10 | -11 | -19 | -4 |
| | Outlet side | In-duct dB | 0 | -6 | -14 | -6 | -8 | -7 | -10 | -12 | -16 | -2 |
| | | Free Field dB | -1 | -14 | -19 | -7 | -8 | -7 | -10 | -12 | -16 | -4 |
| 6 | Inlet side | In-duct dB | -1 | -17 | -16 | -9 | -9 | -8 | -10 | -10 | -17 | -3 |
| | | Free Field dB | -1 | -17 | -21 | -9 | -9 | -8 | -10 | -10 | -17 | -4 |
| | Outlet side | In-duct dB | -1 | -17 | -15 | -8 | -9 | -7 | -10 | -13 | -20 | -6 |
| | | Free Field dB | -1 | -17 | -15 | -8 | -7 | -7 | -10 | -13 | -20 | -7 |
| 7 | Inlet side | In-duct dB | -2 | -9 | -13 | -10 | -8 | -7 | -10 | -12 | -22 | -4 |
| | | Free Field dB | -2 | -17 | -18 | -11 | -8 | -7 | -10 | -12 | -22 | -5 |
| | Outlet side | In-duct dB | 0 | -11 | -12 | -9 | -8 | -9 | -11 | -13 | -19 | -5 |
| | | Free Field dB | -1 | -19 | -17 | -9 | -8 | -9 | -11 | -13 | -19 | -6 |
| 8 | Inlet side | In-duct dB | +1 | -8 | -13 | -9 | -8 | -7 | -10 | -12 | -21 | -4 |
| | | Free Field dB | 0 | -16 | -18 | -10 | -8 | -7 | -10 | -12 | -21 | -4 |
| | Outlet side | In-duct dB | 0 | -5 | -14 | -9 | -8 | -9 | -8 | -12 | -18 | -3 |
| | | Free Field dB | -2 | -13 | -19 | -10 | -8 | -9 | -8 | -12 | -18 | -5 |
| 9 | Inlet side | In-duct dB | -2 | -8 | -13 | -9 | -8 | -7 | -10 | -12 | -20 | -4 |
| | | Free Field dB | -3 | -16 | -18 | -10 | -8 | -7 | -10 | -12 | -20 | -5 |
| | Outlet side | In-duct dB | 0 | -11 | -13 | -9 | -8 | -9 | -10 | -14 | -20 | -5 |
| | | Free Field dB | -1 | -13 | -18 | -10 | -8 | -9 | -10 | -14 | -20 | -7 |

*Elta Fans © 1986*  *92A.86*

*The In-duct Sound Power levels shown on these pages are the result of extensive testing carried out by Elta Fans and by an independent testing authority under the NATLAS accredition scheme. Other sound data is based on known empirical corrections.*

**(b) Zonal sound spectra presentation**

---

## Performance Data

**Elta Fans Ltd**
Wintersells Road, Byfleet, Surrey KT14 7LF, England  ·  Tel: Byfleet (09323) 52021  ·  Telex: 915335 G

**Axial Flow Fan**
**Type 630/150/10**
**1440 REV/MIN**

BS 848 Part 1 1980
Part 2 1985
Type D Installation

### Fan Code
The full fan code uniquely identifies the principle geometry of the fan.

**630/150/10/1440/...**

where —
630 – Nominal diameter. mm.
150 – Hub diameter. mm.
10 – Number of blades.
1440 – Fan speed. rev/min.
... – Blade angle. degs.

### Symbols
$q_v$ – Volume flow
$p_sF$ – Fan Static Pressure
$p_dF$ – Fan Dynamic Pressure (based on Fan Duct area)
$p_tF$ – $(p_sF + p_dF)$ – Fan Total Pressure
$P_R$ – Fan Impeller Power
dBW – Sound Power dB

$$\text{Fan total efficiency \% } = \frac{q_v \times p_tF}{10\,P_R}$$

### Installation
For other installation categories the adjacent relative pressure loss must be added to the required static pressure before using the above graph.

### Motor Data
The motor data shown below is based on manufacturers nominal information. Airstream cooling enables greater than standard motor output to be achieved in some instances depending on the operating ambient. While both direct-on-line and star-delta methods of starting can be used, the currents stated below are based on D.O.L. up to 5.5 kw and star-delta* over 5.5 kw.

**415v/3ph/50Hz supply**

| Blade Angle° | Frame size | Rating kW | Starting amps | Running amps |
|---|---|---|---|---|
| 0-15 | 80b | 0.75 | 8.7 | 1.9 |
| 16-20 | 90S | 1.1 | 14.5 | 2.9 |
| 21-25 | 90L | 1.5 | 19.5 | 3.6 |
| 26-34 | 100La | 2.2 | 27.0 | 4.9 |
| 35-40 | 100Lb | 3.0 | 39.0 | 6.5 |

**240v/1ph/50Hz supply**

| Blade Angle° | Frame size | Rating kW | Starting amps | Running amps |
|---|---|---|---|---|
| 0-15 | 80c | 0.75 | 17.5 | 5.0 |
| 16-20 | 90S | 1.1 | 27.6 | 6.9 |
| 21-25 | 90L | 1.5 | 40.4 | 9.5 |

### Sound Levels
The sound levels shown on the performance curves above are the total Sound Power level In-duct levels on the outlet side of the fan in dB re 1pW. Full sound data relating to Total and Spectra variations for both Inlet and Outlet side, ducted or free-field conditions are shown overleaf.

### Performance Standards
This fan is part of a series of axial flow fans which have been fully tested to the requirements of BS 848 Part 1 (1980) for aerodynamic performance and BS 848 Part 2 (1985) for acoustic performance.

*92A.86   Elta Fans © 1986*

**(a) Airside performance with overall sound power and sound zone presentation**

**Fig 8   Fan performance presentation**

# C401/025

# Effects of ducting configuration on fan sound power levels

**W T W CORY**, CEng, MIMechE, MCIBSE, MIAgrE and **P J HUNNABALL**, MIOA, MSEE
Woods of Colchester Limited, Colchester

BS848:Pt.2:1985 recognises that there can be more than one sound power level associated with a fan operating at a particular duty point. There will be differences in the levels at inlet and outlet for a fully ducted unit. Further differences will be recorded according to installation category. The fan and duct elements act as acoustic impedances so that this is not the end of the story. According to the distribution of ducting between the fan inlet side and its outlet, the fan noise can be affected. This paper describes the results of a series of tests and makes suggestions for optimum placing of fans in their systems.

## 1    INTRODUCTION

For many years it has been recognised that the aerodynamic performance of a fan is dependent on the ductwork connections attached to the fan inlet and/or outlet. If the fan is to develop its maximum pressure capability, then air must be presented to its inlet as a symmetrical and substantially fully developed velocity profile. In like manner, outlet ducting should permit the recovery of excess kinetic energy in the uneven velocity pressure at the discharge plane and its conversion to useful static pressure further along this duct. (Fig.1)

The form of the inlet connection can have a significant bearing on the aerodynamic performance, according to how the fan is ducted. Thus, a spigot may be ideal for a unit attached to its system via a flexible connection. If, however, the fan is unducted, and drawing its air from free space, the spigot will lead to the formation of a 'vena contracta' with corresponding reduction in fan pressure and flow. In such a case, a bellmouth at entry will render any losses negligible.

It is only of recent years that these performance differences have been recognised in a standard method of test (1) and four installation categories defined:

Type A :  free inlet    - free outlet
Type B :  free inlet    - ducted outlet
Type C :  ducted inlet  - free outlet
Type D :  ducted inlet  - ducted outlet

Regretfully many fan manufacturers' catalogues still do not recognise the existence of these differences nor do they indicate the method of test used. Hopefully the wider insistence on certification schemes such as C.A.M.E. (2) will encourage them to provide such information.

In similar manner, fan sound levels used to be considered a fixed quantity dependent only on the position of the operating point on the fan's aerodynamic characteristic (Fig 2). Inlet and outlet sound power levels in open spaces around the fan inlet/outlet were calculated according to classical formulae using end reflection corrections. Research in the 1970's by Baade (3) suggested that this approach was no longer valid but it is only since the publication of a parallel noise measurement standard (4) that differences in fan sound power levels, according to how a unit is ducted, have been recognised by industry. Contrary to the impression given in recent publicity, there is as yet no parallel certification scheme for fan noise levels. It is hoped, however, that this will not be long delayed. Only then will valid comparisons be possible between competing products by the formulation of agreed interpolation methods and a definition of the minimum number of tests to be conducted for any given product range.

## 2    FAN SOUND LEVELS

We now have considerable experimental evidence to support the theory that the sound generated and radiated or transmitted by a fan, are dependent on the acoustic loading at its inlet or outlet. Hence the cross sectional area, length and geometry of any ducting will all have an effect on the sound power levels measured. For the purposes of the British Standard it is, therefore, recognised that with each of the installation categories specified above, there will be a definitive inlet and outlet sound power. Additionally, noise will be radiated from the fan casing, to which will be added the noise from any external motor and transmission. It will thus be seen that there are a number of noise levels that may be specified for any particular flow and rotational speed. But even this is not the end of the story, for Bolton has also shown (5) that outlet in-duct sound power levels measured in an anechoically terminated duct, changed when the open ended inlet duct was altered in length.

Not all researchers in the field (6) are convinced that the differences in these various levels are incapable of resolution. Whilst sound power spectra in the plane wave mode, determined by in-duct tests are invariably higher than those obtained under free field or reverberant room conditions, the differences can usually be explained by the reflection of the sound waves at the fan inlet/outlet when the duct is removed.

The change in acoustic loading on the inlet side due to removal of the anechoic duct leads to a reduced total (i.e. logarithmic addition of inlet and outlet) sound power output of the fan. Such an effect is not thought to be present on the outlet side. Conversely, in the frequency range of higher order modes, in-duct sound power levels have been shown to be lower than those measured under free field conditions. It is thought by some that this may be explained by inaccuracies in the terms for 'modal correction' and 'flow velocity correction' contained in the standard.

An alternative and/or parallel explanation for some of the difference in sound level which have been noted, is the acoustic impedance of the ductwork configuration. Until recently, there were severe practical difficulties in making impedance measurements but these have been reduced with recent advances in digital frequency analysis and correlation techniques. Whereas it was previously necessary to investigate the standing wave patterns by a microphone traverse along the duct for each discrete frequency of interest, it is now possible to use phase matched condenser microphones for simultaneous measurement of sound pressure levels at a known separation. The signals may then be processed through an FFT twin channel frequency analyser to derive impedances from the cross-spectral density function (7) or by a transfer function method (8).

3   ACOUSTIC IMPEDANCE

The specific acoustic impedance i may be defined as the ratio of acoustic pressure p to acoustic particle velocity u and in air is equal to $\rho C$. In a duct, however, this is not a particularly helpful concept and the acoustic impedance I is used, defined as the ratio of acoustic pressure p to the acoustic volume velocity q. With plane wave propagation along a duct of cross-sectional area A and with no reflected waves, then

$$I = \frac{p}{q} = \frac{p}{A_u} = \frac{i}{A}$$

Where reflected waves are present, the pressure and volume velocities are the sum of incident and reflected pressures and the difference between forward and reflected velocities respectively so that the ratio of $\frac{p}{u}$ is generally complex. Knowing the impedance at a point together with the acoustic pressure or volume velocity, it is possible to calculate the unknown parameters. Whilst the main applications of these acoustic impedance concepts have been in reactive silencer design, more recently an impedance model of a ducted fan has been given by Baade (9) where it is considered as a dipole source of noise with internal impedance $I_F$. Acoustic loads of impedance $I_{Li}$ and $I_{Lo}$ are coupled to the end of straight inlet and outlet ducting respectively. Acoustic impedances seen by the fan impeller are $I_i$ and $I_o$. The volume velocities $q_i$ and $q_o$ are equal in magnitude but of opposite sign and are related to the dipole source strength by the equation:

$$q_i = \frac{\Delta p}{I_i + I_F + I_o} = -q_o$$

By manipulation of these terms and noting that the acoustic power flow $w_o = q_o^2 R |I_o|$

$$= \frac{\Delta p^2 R_o}{(I_i + I_F + I_o)^2}$$

Baade deduced that:

$$w_o = \frac{p^2 R \left[ \dfrac{I_{Lo} + j \tan kl_i}{1 + j\, I_{Lo} \tan kl_o} \right]}{\left[ \dfrac{I_{Li} = j \tan kl_i}{1 + j\, I_{Li} \tan kl_i} + I_F + \dfrac{I_{Lo} + j \tan kl_o}{1 + j\, I_{Lo} \tan kl_o} \right]^2}$$

It will be noted that the sound power in the discharge duct is a function not only of the outlet duct length and outlet terminating load, but also of the inlet duct length and inlet terminating load.

Bolton and Margetts (10) have also looked at the influence of changing duct configurations on the noise generated and have concluded that at present, there is no way of estimating the inlet or outlet sound power for one particular installation category from tests carried out on another. Tests are, therefore, necessary in all four categories from which it may be possible to identify those fan designs that are installation sensitive.

4   EXPERIMENTAL PROGRAMME

For any meaningful comparisons to be made between noise tests on fans in a homologous range, and also to compare sound power levels of fans of different types, it is necessary to ensure that the results are accurate and repeatable. They must provide information that can be used by a system designer for noise management and, where necessary, attenuation.

It was determined that for the specific series of experiments for which results will be given, a mixed flow fan (Fig 3a) would be tested with an anechoically terminated inlet and/or outlet, so that the various installation categories could be replicated. A bifurcated axial flow fan (Fig 3b) would also be tested for category D, but with the distribution of loading altered between the inlet and outlet such that the overall fan pressure remained virtually constant for the rated flow.

Bolton (11) has noted that there are severe problems with the sampling tube (turbulence screen) whilst as previously mentioned, there are doubts as to the value of the modal and flow corrections. It was, therefore, determined that the microphone would be shielded in the experiments by a simple polyurethane foam ball which has no limitations on its use as to swirl. This may still be present at flowrates below the point of maximum efficiency, a toroid being produced upstream of the fan as the air accelerates on to the impeller blades. The anechoic terminator was of the "stepped" type consisting of a number of parallel acoustically lined duct sections of varying diameters bolted together. This type of terminator has a more than adequate performance and is easy to manufacture. It requires a minimum of storage space as the steps can be stored inside each other. The steps were also chosen to correspond to a series of test duct sizes, so that one set of parts can cover all sizes in a whole series of fans.

## 5 APPARATUS AND TESTS

Readings were taken at constant speed and a flowrate approximating to the point of maximum efficiency on the aerodynamic characteristic. The flow was maintained for each installation category and ductwork configuration so that comparisons could easily be made. Sound power levels were recorded for both the inlet and outlet (in-duct or free field as appropriate) in $\frac{1}{3}$ octave bands at centre frequencies from 31.5Hz to 16kHz. This enabled pure tones to be more easily identified and also produced more data for the analysis. An Ono Sokki type CF910 dual channel FFT analyser was used so that the data captured could be held on floppy disc. Each of the experimental ductwork sets is shown in Fig.4 and the point of aerodynamic loading will be noted.

Whilst the uncertainty of readings below 50Hz would exceed 3.5dB, which at 95% confidence limits and a normal distribution of data should be multiplied by 2, it was, nevertheless, felt essential to include them. These low bands may encompass the rotational frequency and sometimes even its multiples. The sub-multiples of the blade passing frequency (which may be of more importance at very low flowrates when the impeller blades are stalled), will also be present. Noise in these bands often makes a considerable contribution to the overall or linear sound power level, more especially with the centrifugal fan types, but still significant with the mixed flow fan used in the first series of these experiments and even with many axial flow fans.

## 6 RESULTS

Readings for the Mixed Flow Fan have been plotted as inlet and outlet sound power levels in $\frac{1}{3}$ octave bands (full lines) and octave bands (dashed lines) for all Installation Categories (Fig 5). Where a duct is present then figures are, of course, in-duct levels. Open inlet and/or outlet figures are the result of an hemispherical integration in accordance with BS848: Part 2:1985.

With the Bifurcated Axial Flow Fan, similar plots have been completed (Fig 6) but as these tests were only conducted for Category D, the analyses are restricted accordingly. They show only the effects of inlet majority loading versus outlet majority loading.

## 7 CONCLUSIONS

The tests confirm that the sound power levels of fans vary according to the Installation Category. Differences for those tested are significant, but cannot be explained solely by end reflection considerations. It will, therefore, be necessary for appreciably more tests to be conducted in the future, if meaningful advice is to be given to

the system designer.

How the fan is loaded is also seen to be of importance and perhaps the Category D tests in BS848:Part 2 should reflect this. In passing, it will be noted that the axial fan is seen to generate considerable quantities of low frequency noise. This is our experience over a number of tests. It is contrary to the generalized spectra given in most text books, where the noise is shown to peak at blade passing frequency. Perhaps monopole generation as determined by volume displacement is of more importance than the Dipole Sources such as vortex shedding from the blades, considered to be dominant. It may, however, reflect the fact that only with the introduction of the 1985 test code, has noise at 63Hz received sufficient consideration.

Where the suppression of low frequency noise is important, the results indicate that for the particular fan tested, distribution of the ducting equally on either side will probably give the lowest results. Above the blade passing frequency, outlet noise is invariably higher than inlet noise. Whilst relatively easy to attenuate, a 'blowing' duty for the fan is indicated.

It will be noted that for both Mixed and Axial Flow fans, the noise at rotational frequency is significant. Although the fans were balanced to G2.5 or better, this can perhaps be explained by the rigid coupling of the fans to the test ducting, as demanded by the test standard. The resultant conversion of vibration into sound according to the duct radiation efficiency may be of importance. This suggests that the fan should be coupled to the test ducting by flexible connections. Perhaps, however, this will only introduce new errors as, unless these connections are taut and axially aligned, blade tips of axial and shroudless mixed flow fans could be 'starved' and a further source of noise introduced.

REFERENCES

(1) BRITISH STANDARDS INSTITUTION Fans for General Purposes Part 1 Methods of Testing Performance - BS848:Part 1, 1980.

(2) BRITISH STANDARDS INSTITUTION Certification of Air Moving Equipment - Part 1 Fan Rating Procedures, 1986.

(3) BAADE, P. Effects of acoustic loading on Axial Flow Fan noise generation - Noise Control Engineering, 1977.

(4) BRITISH STANDARDS INSTITUTION Fans for General Purposes Part 2 Methods of Noise Testing - BS848:Part 2, 1985.

(5) BOLTON, A. N. A new fan noise measurement standard BS848:Part 2:1985 - Proceedings of the Air Movement and Distribution Conference - Purdue University, Indiana, 1986.

(6) NEISE, W., HOLSTE, F., HOPPE, G. Experimental comparison of standardized sound power measurement procedures for fans - proceedings Inter-noise, 1988.

(7) SEYBERT, A.F., ROSS, D. F. Experimental determination of acoustic properties using a two-microphone random excitation technique - Journal of Acoustics Society of America, 1977.

(8) CHUNG, J. Y., BLASER, D. A. Transfer function method of measuring in-duct acoustic properties - Journal of Acoustics Society of America, 1980.

(9) BOLTON, A. N., MARGETTS, E. J. The influence of impedance on fan sound power - Paper C124/84 Conference on Installation Effects in Ducted Fan Systems (I.Mech.E.), 1984.

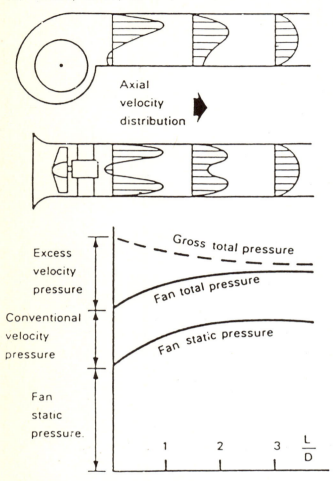

Fig 1   Static pressure regain downstream of a fan

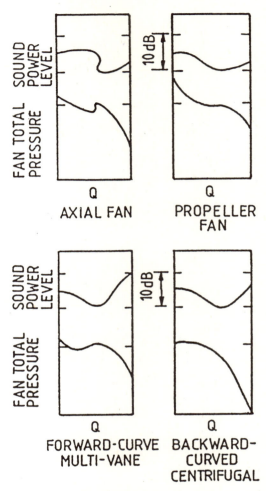

Fig 2   Typical shape of sound power level characteristics

Fig 3   Test fans

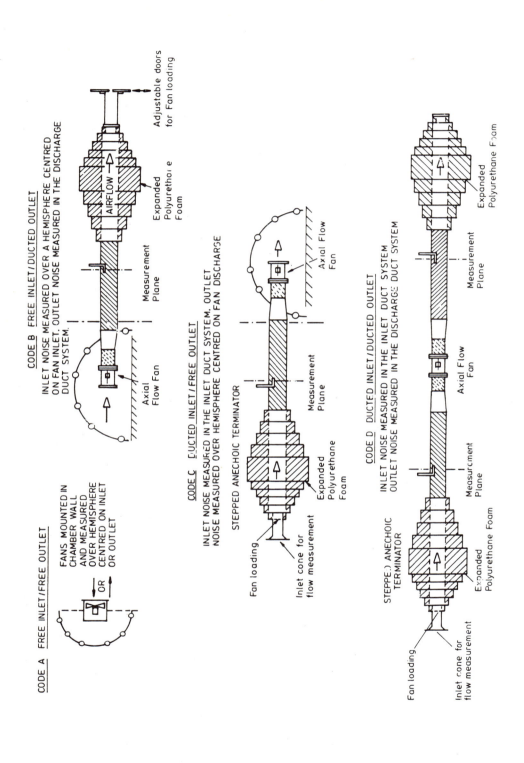

**CODE A   FREE INLET/FREE OUTLET**

FANS MOUNTED IN
CHAMBER WALL
AND MEASURED
OVER HEMISPHERE
CENTRED ON INLET
OR OUTLET

**CODE B   FREE INLET/DUCTED OUTLET**

INLET NOISE MEASURED OVER A HEMISPHERE CENTRED
ON FAN INLET. OUTLET NOISE MEASURED IN THE DISCHARGE
DUCT SYSTEM.

AIRFLOW

Adjustable doors
for Fan loading

Expanded
Polyurethane
Foam

Measurement
Plane

Axial
Flow Fan

**CODE C   DUCTED INLET/FREE OUTLET**

INLET NOISE MEASURED IN THE INLET DUCT SYSTEM. OUTLET
NOISE MEASURED OVER HEMISPHERE CENTRED ON FAN DISCHARGE

STEPPED ANECHOIC TERMINATOR

Axial Flow
Fan

Measurement
Plane

Expanded
Polyurethane
Foam

Fan loading

Inlet cone for
flow measurement

**CODE D   DUCTED INLET/DUCTED OUTLET**

INLET NOISE MEASURED IN THE INLET DUCT SYSTEM
OUTLET NOISE MEASURED IN THE DISCHARGE DUCT SYSTEM

Expanded
Polyurethane Foam

Measurement
Plane

Axial Flow
Fan

Measurement
Plane

STEPPED ANECHOIC
TERMINATOR

Expanded
Polyurethane Foam

Fan loading

Inlet cone for
flow measurement

**Fig 4    Arrangement of test ducting for measurement of fan in-duct and
free field sound power levels**

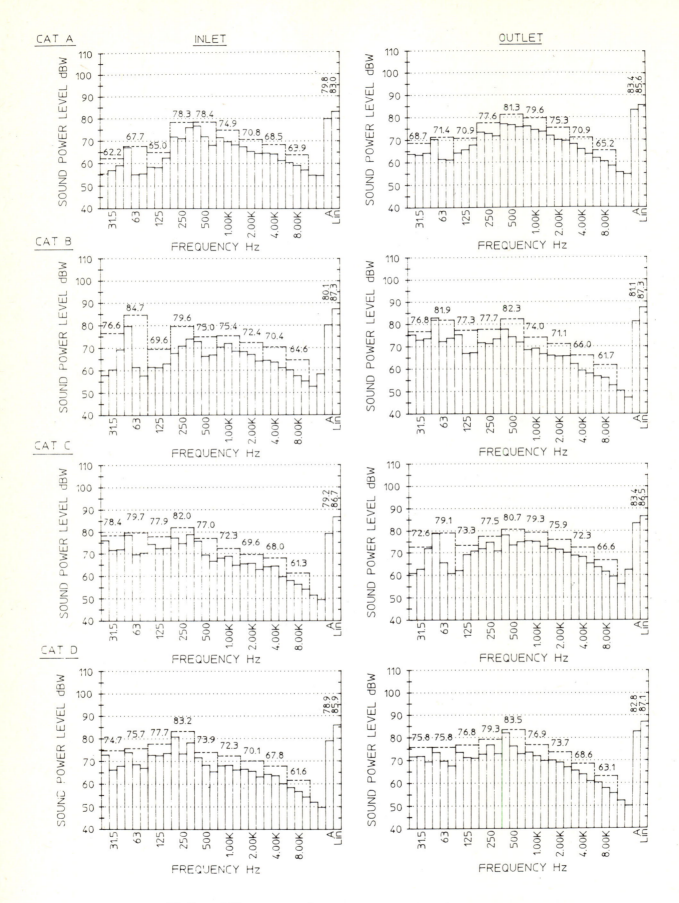

Fig 5    315 mm Axcent 2 mixed flow fan at 2850 r/min and maximum efficiency (0.53—0.54 m³/s)

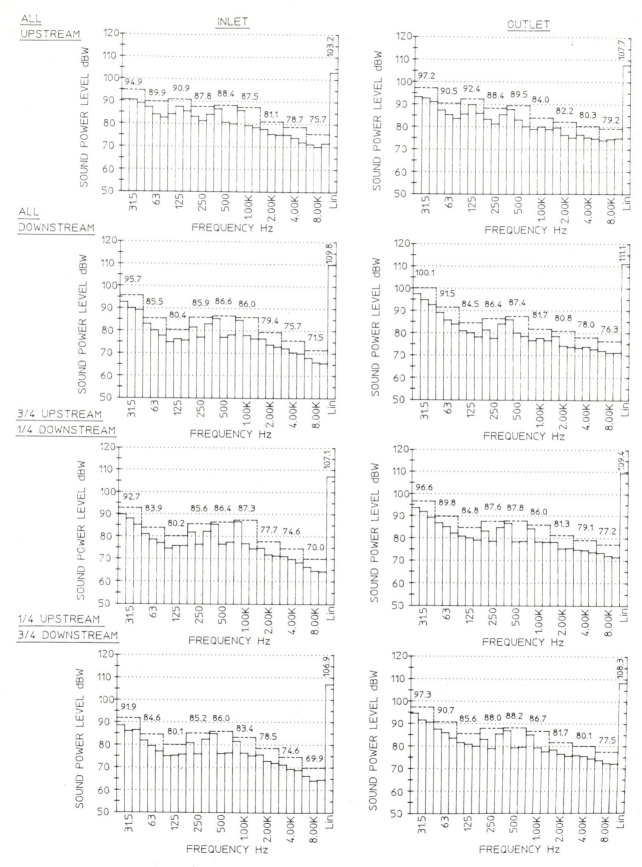

Fig 6    Variation in sound power levels for 610 mm bifurcated axial flow
         fan 18.5 kW 2920 r/min installation category D according to
         distribution of aerodynamic loading

# Installation effects on fan noise

**W NEISE**, Dr-Ing
Deutsche Forschungsanstalt fur Luft und Raumfahrt, Berlin

SYNOPSIS

The present paper deals with excess fan noise that is generated by installation effects. Three different types of installation effects are discussed: Inlet flow conditions, acoustic loading and fan operation control. Among the examples given are inlet boxes, duct bends and flow diffusers mounted upstream of centrifugal and axial fans; volume flow control by means of throttling and inlet guide vanes; change in total fan sound power when operating with free inlet/ducted outlet, ducted inlet/free outlet, and ducted inlet/ducted outlet.

## 1 INTRODUCTION

The primary cause of the aerodynamic noise from low to medium speed industrial fans are the steady and unsteady forces that are exerted by the flow on the fan blades and vanes. Unsteady forces are generated when a fan rotor operates in a spatially non-uniform flow field, because magnitude and angle of attack of the flow as seen by the blades change with angular position. Steady flow distortions lead to periodic blade forces and tonal noise components, and stochastic variations of the flow field produce random blade forces and in turn random noise components.

Fan noise control at the source means avoiding or reducing as many of the various noise generation mechanisms as possible, i.e., diminishing the fluctuating forces on blades and vanes. Many studies have been published in the past in which methods were developed to design quiet fans; for a review of these see reference (1). To achieve the lowest possible noise level in a fan installation, it is also necessary to select the optimum fan type for the given aerodynamic task. In a recent paper by the author (2) it was shown that the different fan types vary significantly in their specific sound power level which is the fan sound power level reduced in a certain way by the fan aerodynamic duty. For example, centrifugal fans with backward curved blades proved quieter in this comparison than axial fans. In many applications a given flow rate Q has to be pumped through a given duct system which requires a certain total fan pressure $\Delta p_t$. In this case there is generally only a limited choice of fan types that are suitable for the given task. However, there are also technical problems which can be solved with different combinations of Q and $\Delta p_t$, and in these cases the fan duct system should be layed out in a way which allows employment of low noise fan types. As an example, a heat exchanger for a given heat transfer rate can be designed to have a large flow area and a small depth or, *vice versa*, a small flow area and a large depth. In the first case, the cooling fan would have to deliver a large flow rate at a low flow resistance, and in the second case a low flow rate against a high back pressure.

In practical fan installations often unexpected fan noise problems arise although the fan is acoustically well designed and appropriately selected. In this paper the problem of excess fan noise is addressed that is generated by the fan installation. Three different types of installation effects are discussed: Inlet flow conditions, acoustic loading and fan operation control.

## 2 EFFECTS OF INFLOW CONDITIONS ON FAN NOISE

Inlet flow distortions and incident turbulence generate unsteady forces on the rotating impeller blades and, hence, generate noise. Such non-uniform flow conditions are caused by various duct elements like inlet boxes, bends, corners, diffusers, throttling devices, guide vanes, etc.

Schmidt (3) investigated the effect of a 90° duct bend mounted at various axial distances upstream of an axial fan on the radiated sound pressure spectrum (Figure 1). As the result of secondary flows and flow separation in the duct bend, the velocity profile behind the bend is not axisymmetric any more, and the turbulence intensity is increased. At short distances between duct bend and axial fan, the low frequency random noise components are increased significantly, and the same is true for the blade passage frequency component which lies in the 400 Hz one-third octave band. The most critical range for the axial distance x/D is 1.5 to 4 (D = impeller diameter). Schmidt concluded that a minimum distance of eight duct diameters is necessary to avoid excess noise generation. It is felt here that this minimum distance is also a function of the radius of curvature of the bend which was not reported.

In many practical centrifugal fan installations, an inlet box is mounted at the fan intake which involves a 90°-turn of the flow directly before entering the impeller. This situation is similar to a 90° duct bend upstream of the fan inlet. Neise and Barsikow (4) showed that the presence of the inlet box increases the noise output of the fan by up to 4.5 dB(A) as compared to an axisymmetric inlet nozzle, see Figure 2.

This result was obtained with a set of fully opened inlet guide vanes at the fan inlet in both cases. The sound pressure spectra measured in the anechoically terminated fan inlet duct show that due to the distorted velocity profile at the entrance plane of the impeller, the tonal as well as the random noise components are increased, however, this effect is not as strong as in the case of the axial fan, compare Figure 1.

In another paper (5), a centrifugal fan with an inlet box but without inlet guide vanes was used to study the effect of the inflow conditions on the noise generated. It was found that the presence of the inlet box in the absence of inlet guide control means leads to unsteady fan operation due to steady and unsteady swirl components generated in the inlet box. This swirl can be eliminated by a short splitter plate that extends from the impeller shaft to the bottom of the inlet box. As a result, operation of the fan becomes smooth and steady, and the aerodynamic and acoustic fan performance are improved, see Figure 3. The non-dimensional parameters used are defined as follows.

$$\varphi = Q/\pi bDU \qquad \text{(flow coefficient)} \qquad (1)$$

$$\psi = 2\Delta p_t/\varrho_0 U^2 \qquad \text{(pressure coefficient)} \qquad (2)$$

$$\eta = Q \cdot \Delta p_t/P_{el} \qquad \text{(efficiency)} \qquad (3)$$

Here Q and $\Delta p_t$ are volume flow and total pressure head of the fan. Width, diameter and tip speed of the impeller are denoted by b, D and U, respectively. The density of air is $\varrho_0$. Note that since only the electric power input $P_{el}$ to the drive motor was measured, the fan efficiency as defined in equation (3) includes the losses in the DC-motor, pulley drive, bearings and the fan.

The presence of the splitter plate improves the fan characteristics mainly at high volume flows. The maximum efficiency is raised from 0.63 to 0.68, the corresponding pressure coefficient by about 4 per cent . At the same time the noise radiated is decreased over a wide range of fan operating conditions with the largest reductions occurring in the outlet duct. The level of the blade passage frequency which represents the most annoying fan noise component is diminished by up to 14 dB.

Figure 3 also shows sound pressure spectra measured at 3000/min and $\varphi=0.145$. The addition of the splitter plate lowers the frequency spectra on both inlet and outlet over the entire frequency range.

To explain the experimental observations described above, hot wire measurements were made at various locations in the inlet box. The results indicated that without the splitter plate, the flow is characterized by strong swirl components and turbulence levels up to 31 per cent. With the splitter, the swirl is removed and at the same time the turbulence levels are reduced by about a factor of 3.

Another example for the effect of inflow conditions on fan noise is given in Figure 4 (after references (6 and 7)). There, sound power spectra measured in the anechoic inlet and outlet ducts of an axial and a centrifugal fan are shown for two cases: In the first, a straight duct of 600 mm internal diameter is connected to the fan

intake, and in the second case a 650 mm long diffusor from 500 to 600 mm internal diameter (10° enclosed angle). Diffusers are known to produce a velocity profile with a pronounced maximum on the axis together with high turbulence intensities near the wall. While there is only a very small effect visible in the spectra of the centrifugal fan, the sound power level of the axial fan is increased over almost the entire frequency range in both inlet and outlet duct, resulting in a by up to 4.5 dB higher A-weighted sound power level.

The two example discussed above clearly show that axial fans are more sensitive to distorted inlet flow conditions than centrifugal fans.

## 3   EFFECTS OF FAN OPERATION CONTROL ON FAN NOISE

In practical fan installations it is often required to vary the volume flow over a fairly wide range. It was shown by Neise and Barsikow (4) that it is advisable to do this by controlling the fan speed rather than by throttling or by inlet guide vanes, because that is advantageous not only with respect to energy consumption but also from an acoustic point of view. In Figure 5 linear and A-weighted sound pressure level in the anechoic inlet duct of a centrifugal fan are plotted as functions of the flow rate for the three volume flow control methods mentioned above. While for throttle and inlet guide vane control, the sound pressure level radiated increases with decreasing flow rate, it is sharply reduced in the case of impeller speed control. If a volume flow of half the nominal rate is considered, the linear and A-weighted noise levels are by 8-10 dB and 20 dB(A), respectively, lower in case of fan speed control than with throttling or vane control.

## 4   ACOUSTIC LOADING EFFECTS ON FAN NOISE

It was shown by Cremer (8), Baade (9) and Wollherr (10) that the sound power generated by a fan is not only a function of its impeller speed and operating condition, but it also depends on the acoustic impedances of the duct systems connected to its inlet and outlet. Therefore, fan and duct system should be matched not only for aerodynamic reasons but also because of acoustic considerations. Care has to be taken that tonal fan noise components do not coincide with resonance frequencies of the duct system.

Baade (9) stated that a very noticeable increase in blade tone level occurs whenever the effective length of a fan duct is made to be equal to half the sound wavelength. On the other hand, a lower noise level can be obtained if a tone falls on an anti-resonance of the system. Bommes (11) reported a case where by appropriate choice of fan diameter, speed and blade number, the blade tone was placed on a frequency where the radiation efficiency of the fan was minimum. The resultant level reduction was 25 dB. Wollherr (10) showed that a substantial change in the blade passing frequency level can be obtained by mismatching the acoustic impedances of the fan (source) and the duct system (load). In his example, a 17 dB drop in the tone level in the fan inlet duct was achieved by manipulating the length of the outlet duct.

Acoustic loading effects became visible in an experimental study by Neise, Holste and Hoppe (12) in which the sound power of six fans of different designs was measured using three different standardized measurement procedures, i.e., the free-field method (ISO 3744 or DIN 45635 Teil 1), the reverberation room method (ISO 3741/3742 or DIN 45635 Teil 2), and the in-duct method (ISO/DIS 5136.2 or DIN 45635 Teil 9). The first two methods are to be applied to determine the sound power radiated from the fan inlet or outlet into free space, and the in-duct method yields the sound power radiated into a duct. Only the data from the free-field measurements and the in-duct measurements are discussed here.

The free-field measurements were made once with the fan inlet or outlet, respectively, radiating directly into open space and a second time with a cone connected to the inlet or outlet side. The cone of 2.1 m length and 0.5 to 1.6 m diameter was meant to generate an acoustic load impedance similar to that of an anechoic duct. For all methods, simultaneous sound measurements were made in the duct remaining on the opposite side of the fan.

Figure 6 shows the results for one of the test fans, a centrifugal fan with backward curved blades. The top diagram of Figure 6a depicts the sound power spectra measured on the fan inlet side using the two measurement codes described above. The diagram in the middle shows the spectra measured in the anechoic outlet duct for the three inlet configurations, and the bottom one gives the total sound power emitted, i.e., the sum of inlet and outlet noise. Figure 6b is analogous to Figure 6a, only that the roles of inlet and outlet are interchanged; the comparison of in-duct levels and free-field levels is now shown in the diagram in the middle.

In the frequency region of plane wave propagation in the fan duct (here: f < 400 Hz for d = 0,5 m), the in-duct power levels are higher than the free-field levels measured without cone. The reason for this is that part of the sound energy is reflected at the fan opening. In case of the fan outlet being open, the end reflection results in increased levels in the fan inlet duct, and as a result the total fan sound power is not changed (see the bottom diagram of Figure 6b) but merely shifted from outlet to inlet. The situation is somewhat different when the inlet is open (Figure 6a). The end reflection is again manifested in the different levels obtained in the duct and in free space, but there is no increase in the fan outlet duct, and accordingly the total sound power is lower than with the inlet ducted. Hence, a change in acoustic loading on the inlet side has an influence on the outlet duct sound power and the total sound power, while such an effect is not present when the acoustic impedance is varied on the outlet side.

The addition of the cone provides an acoustic load impedance similar to that of the anechoic duct, and consequently the sound power levels measured in the free-field and in the duct agree quite well, in particular at the frequencies, where the reflection coefficient of the cone is low, i.e., at 50 Hz and above 100 Hz.

The acoustic loading effects as shown in Figure 6 are representative for all fans measured in reference (12). In case of very large fans,

acoustic loading are less important, because they take place only in the frequency range of plane wave propagation with respect to the cross dimensions of fan inlet or outlet, i.e., at very low frequencies.

## 5 CONCLUSIONS

Experimental results from published literature are presented to show the excess fan noise produced by installation effects. Three different types of installation effects are discussed: Inlet flow distortions, fan operation control and acoustic loading.

Examples are given for the noise of a centrifugal fan with and without inlet box, the effect of unsteady swirl flow in the entrance plane of a centrifugal impeller, the influence of an upstream duct bend upon the noise generated by an axial fan, flow diffusers mounted upstream of centrifugal as well as axial fans, and excess noise as a result of fan volume flow control by means of throttling and inlet guide vanes.

The experimental results reveal that axial fans are more sensitive to disturbed inflow conditions with regard to the radiated noise than centrifugal fans.

Experimental date are also presented for the change in total fan sound power when operating in type B, C, and D installations, i.e., with free inlet/ducted outlet, with ducted inlet/free outlet, and with ducted inlet/ducted outlet, respectively. Quietest fan operation is achieved when the fan inlet is not ducted.

## REFERENCES

(1) Neise, W. Fan noise – generation mechanisms and control methods. Internoise, Avignon, France, 30.8-1.9.1988, pp. 767-776.

(2) Neise, W. Geräuschvergleich von Ventilatoren. Die spezifische Schalleistung zur Beurteilung des Geräusches unterschiedlicher Ventilatortypen. *Heizung Lüftung Haustechnik* **39** (1988), pp. 392-399.

(3) Schmidt, L. Der Einfluß eines 90°-Rohrkrümmers auf die Schallemission eines Axialventilators. *Luft- und Kältetechnik* (1976), pp. 287-290.

(4) Barsikow, B., and Neise, W. Der Einfluß ungleichmäßiger Zuströmung auf das Geräusch von Radialventilatoren. *Fortschritte der Akustik, DAGA'78*, Bochum. VDE-Verlag Berlin (1978), pp. 411-414.

(5) Neise, W. and Hoppe, G. Effect of inflow conditions on centrifugal fan noise. First European Symposium on Air Conditioning and Refrigeration, Brussels Belgium, 5-6 November 1986, pp. 165-172.

(6) Neise, W., Hoppe, G. and Herrmann, I.W. Intercomparison of fan noise measurements using the in-duct method ISO/DIS 5136. Internal report DFVLR-IB 22214-86/B1 (1986).

(7) Neise, W., Hoppe, G. and Herrmann, I.W. Geräuschmessungen an Ventilatoren – Verglei-

chende Untersuchungen nach dem Kanal-Verfahren DIN 45 635 Teil 9 bzw. ISO/DIS 5136. *Heizung Lüftung Haustechnik* **38** (1987), pp. 343-351

(8)   Cremer, L. The treatment of fans as black boxes. *J. Sound Vibr.* **16** (1971), pp. 1-15.

(9)   Baade, P.K. Effects of acoustic loading on axial flow fan noise generation. *Noise Control Engineering* **8** (1977), 5-15.

(10)   Wollherr, H. Akustische Untersuchungen an Radialventilatoren unter Verwendung der Vierpoltheorie: Doctoral Dissertation Techn. Univ. Berlin (1973).

(11)   Bommes, L. Lärmminderung bei einem Radialventilator kleiner Schnelläufigkeit unter Berücksichtigung von Zungenform, Zungenabstand und Schaufelzahl. *Heizung Lüftung Haustechnik* **31** (1980), pp. 173-179 and 210-218.

(12)   Neise, W., Holste, F. and Hoppe, G. Experimental comparison of standardized sound power measurement procedures for fans. Internoise, Avignon, France, 30.8-1.9.1988, pp. 1097-1100.

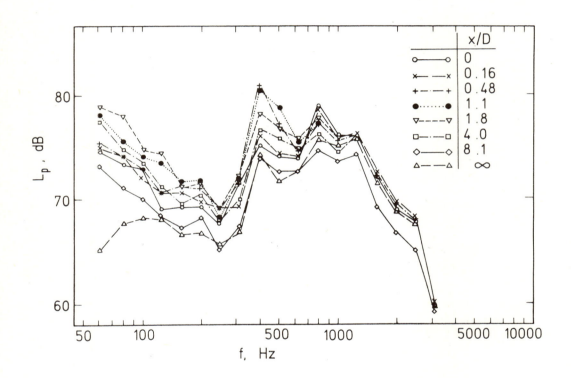

Fig 1      Sound pressure spectra of an axial fan with a 90° duct bend mounted at various distances *x* from the fan inlet [after Schmidt (**3**)]

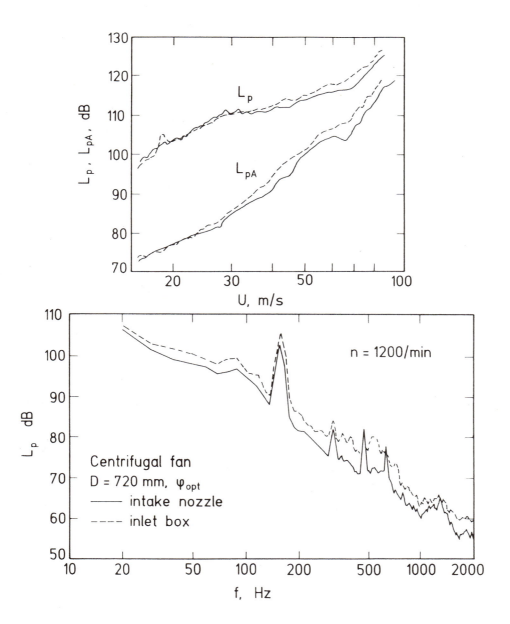

Fig 2    Influence of an inlet box on the noise level of a centrifugal fan
         [after Neise and Barsikow (**4**)]

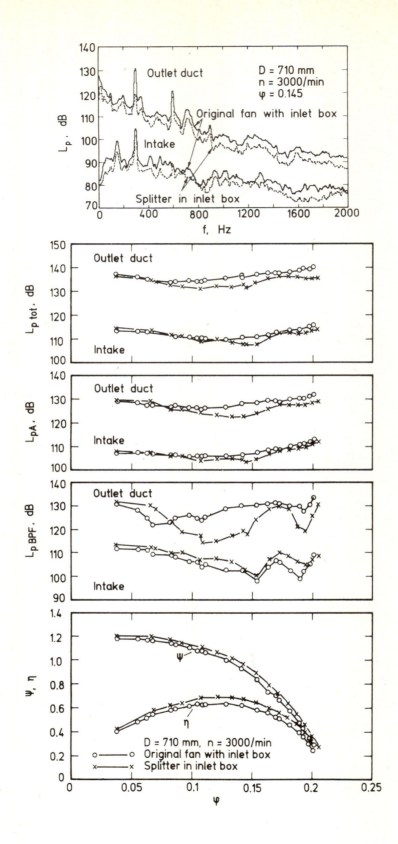

Fig 3    Effect of a splitter plate in the inlet box of a
         centrifugal fan on the acoustic and aerodynamic
         fan performance curves and on the sound
         pressure spectrum. n = 3000/min [after Neise
         and Hoppe (5)]

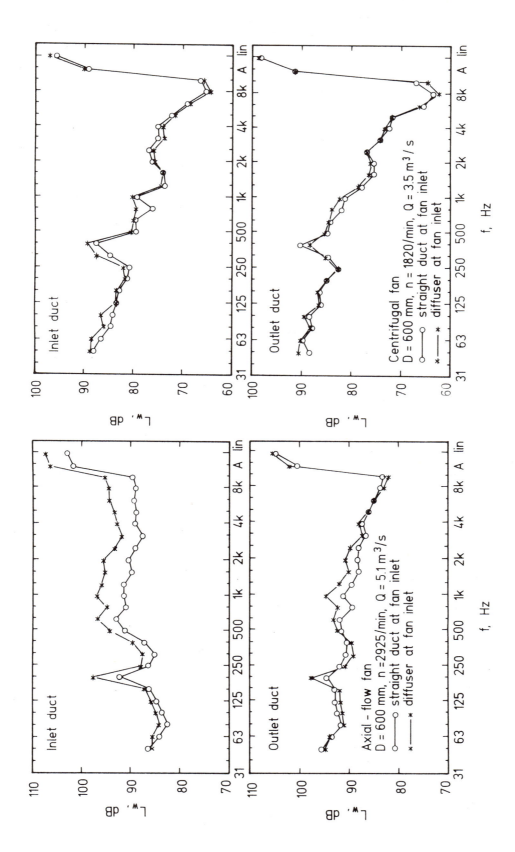

Fig 4 Influence of a diffuser at the intake of an axial and a centrifugal fan on the sound power spectra [after Neise *et al* (**6, 7**)]

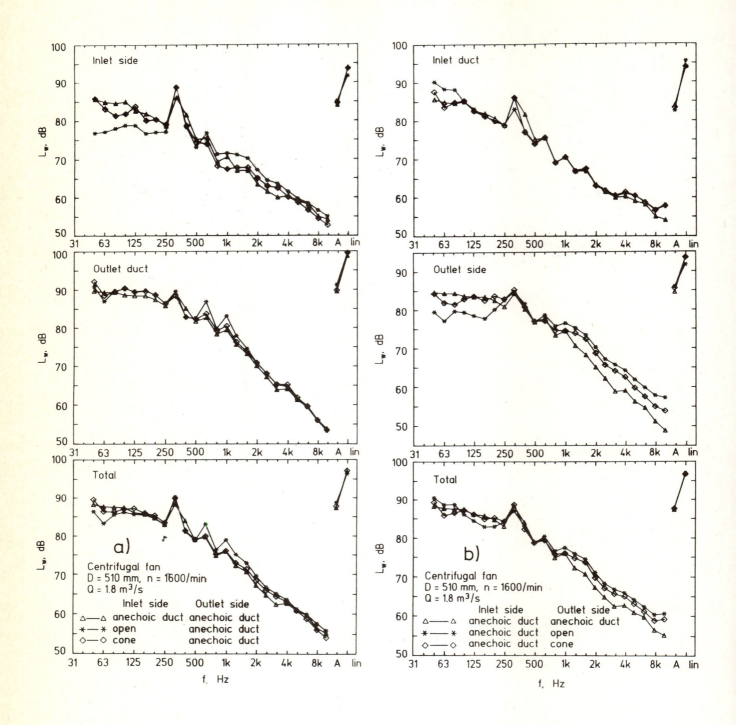

Fig 5    Effect of volume flow control of a centrifugal fan on the sound pressure level generated in the fan inlet duct [after Neise and Barsikow (4)]

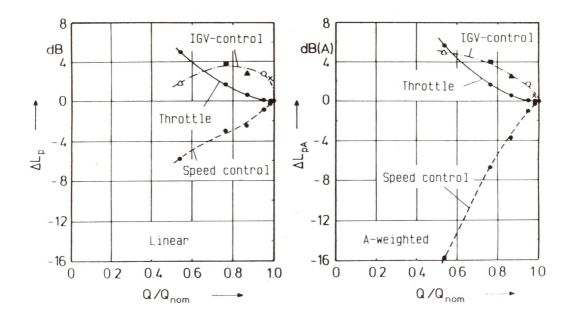

Fig 6    Effects of acoustic loading on the sound power level of a centrifugal fan
         with backward curved blades (a) for different inlet side configurations
         (b) for different outlet side configurations [after Neise *et al* (**12**)]

# C401/003

# Noise reduction and fan design for dust extraction equipment

**G M CHAPMAN** , BSc, PhD, CEng, FIMechE
Department of Mechanical Engineering, University of Technology, Loughborough
**K S SMITH**, BSc, AMIMechE
DCE Group Limited, Thurmaston, Leicester

SYNOPSIS  Methods for noise emission reduction for dust extraction plant are presented. The development of acoustic hoods to lower the noise levels by 10 dB(A) are discussed and the use of modern vibration analysis techniques to assist with design are reviewed.

## 1    INTRODUCTION

Conventional dust extraction equipment employed in commercial environments relies upon fans drawing the air flow through filter elements. The fans are usually fitted down stream of the filter the allow collection of the particles in the air stream without contamination of the fan impeller. Consequently fan generated noise is introduced to both the air stream and the exhaust ducting.

International legislation to limit personnel exposure to noise levels varies considerably throughout the world, although there is an increasing trend for countries to seek lower and safer exposure limits. Manufacturers of equipment have to meet the individual demands of each country, whilst attempting to produce commercially viable standard products suitable for the most stringent criteria.

Recognising the constraints between commercial acceptability and the desirability of low noise emission, a joint programme was set up under the sponsorship of the Teaching Company Scheme introduced by SERC and DTI. The Associate employed on the contract was set the objective to reduce the emitted noise from a specific dust extraction product range down to 85 dB(A). This objective presented a severe challenge as the current equipment typically had a noise emission in the range 90 to 100 dB(A) depending upon the air flow volume.

## 2    ESTABLISHING REFERENCE DATA

At the outset of the project it was decided to establish a reference of the noise emitted from all the units. A standard test procedure was developed using point measurement of sound pressure energy at a defined fixed distance (1 metre) from the the rear of the fan drive motor for the standard configuration. The measurement location was dictated by the construction of the dust extraction range as the fully assembled structure did not permit free access around the fan unit and the location chosen was considered to give noise level results similar to those measured out on site. A check was undertaken to ensure that the noise levels measured at this point were representative of the true noise emission from the assembly. A typical dust extraction unit is shown in Fig 1. The data obtained for these preliminary tests is shown in the first column of Table 1. The results were used as a qualitative assessment of noise emission rather than as a test under a recognised measurement standard.

In the conventional extraction system currently used, the air is drawn into the centre of a radial impeller and exhausted vertically downwards directly into the local environment. This arrangement is used so that units mounted outdoors do not allow rain etc, to enter the system through the exhaust ducting. For many applications it has been possible to site the equipment in a locality away from personnel and the noise emission has not been considered to be a hazard. However an increasing number of applications require the extractor units to be installed either in the vicinity of the workforce or in an area where noise emission is unacceptable to environment. To meet these requirements a range of splitter attenuators are offered but they carry the penalty of a requirement for more space and a possible extra fabrication cost for a support structure.

## 3    INITIAL TESTS

The fan is cantilever mounted directly onto the drive shaft of an electric motor and is supported on a fabricated structure. This structure also includes an integral header unit through which the air stream is drawn into the fan. Noise can be transmitted directly to the casing via the motor mounting and the fan casing into the structure via direct mechanical coupling at the motor and fan case mounting points. However the most significant noise emission derives from noise excitation carried in the gas stream deriving from blade pass frequency. Initially noise reduction techniques concentrated on absorption of noise emission from the outlet air stream.

# 4 ACOUSTIC HOOD DESIGN

It was initially decided to undertake all the preliminary tests on the fan/motor system with the worst noise emission history in the production range. The particular assembly tested was in fact from the middle of the range where the air flow capability was reasonably high for a structure which was comparatively flexible. Larger fan assemblies are produced using thick metal plate and smaller fans move less air volume. All tests were conducted using unbaffled outlets and the fans were run at their typical operating point for free discharge.

## 4.1 Simple acoustic enclosure

The fan assembly was enclosed in an acoustic box and the enclosure internal surfaces were lined with a single 20mm layer of acoustic foam. The acoustic foam used was a fire resistant poly-ether based material with a density of 30 kg/m$^3$. The enclosure design was dictated by the requirement that it should not occupy a significantly larger volume than the un-enclosed motor/fan assembly. The bounds of the enclosure are illustrated in Fig 2.

To permit air exhaust from the enclosure whilst attempting to reduce the air borne noise, the maximum possible air flow path length was achieved by placing the exhaust port on the side panel diametrically opposite the fan casing exhaust. The configuration for fan orientation and exhaust port location are shown in Fig 3.1.

With such a configuration it was found possible to reduce the noise emission from 100 dB(A) down to 87 dB(A) with a minimal loss in fan performance. However the introduction of the acoustic enclosure caused a low frequency rumbling effect to be stimulated in the coupled Header; possibly initiated by the air flow setting up standing wave vibration excitation in the Header unit and thus exciting panel vibrations. To cure this excitation it was necessary to incorporate a stiffening element on one of the panels.

Grilles mounted on the side of the enclosure were not acceptable to the design philosophy of the extractor unit as the possible ingress of rain water etc, was considered to be detrimental to the life of the system. The only solution to this problem would have been to develop weather cowls with the additional problems of throttling the exhaust flow and increasing the cost of manufacture of the units.

## 4.2 Acoustic enclosure with vertical exhaust

Attention was therefore turned to arranging for the exhaust grille to be situated in the base panel of the acoustic enclosure at the opposite end to the fan exhaust as illustrated in Fig 3.2. A noise emission level of 85 dB(A) was achieved even though the air stream path length was almost a minimum. The effect upon fan performance was minimal but it was felt that a greater noise emission reduction should be targeted in order that similar designs of acoustic enclosures for the larger assemblies would result in similar noise level reductions.

## 4.3 Introduction of full baffle

In an attempt to increase the air flow path a full baffle was introduced (Fig 3.3) to force the air circulation around the whole enclosure before exhausting to the environment. The emitted noise level was dropped to 80 dB(A) but at the expense of appreciable loss of fan performance. The extractor system was reduced to 75% of its normal un-enclosed value.

## 4.4 Acoustic enclosure with half baffle

To reduce the throttling process, the baffle was reduced to half height (Fig 3.4) so that some air flow was possible direct from fan exhaust to outlet grille. This resulted in restoration of fan performance but raised the noise emission to 88 dB(A); a value greater than for a similar system without a baffle. It was assumed that the partial baffle raised the noise emission level as a consequence of the introduction of vortices in the air stream and so no attempts were made to investigate other partial baffle geometries.

## 4.5 Re-orientation of fan

At this stage in the investigation it was accepted that lined acoustic enclosures offered potential noise reduction possibilities which were likely to reach the objective of 85 dB(A) for the whole range, so long as a method could be found to ensure a sufficient air stream flow path inside the enclosure whilst not effecting the fan performance. It was also felt desirable to maintain the Company policy of exhausting vertically downwards to enable the extraction systems to be sited outside of users premises. Baffles were not considered to be desirable and so the decision was taken to mount the fan in such a way that the fan exhaust was diametrically opposite the exhaust grille. This necessitated the need to rotate the fan anti-clockwise by 90$^o$ to the normal configuration (Fig 3.5). The measured noise emission level for this configuration was dropped to 83 dB(A) and it was decided to adopt this configuration for the whole fan range.

## 4.6 Noise reduction for complete fan range

For all designs a minimum gap between fan casing outlet and the adjacent wall was maintained to minimise the throttling effect. The results obtained are reported in Table 1, where the initial un-enclosed system is compared to the final design for all seven current fan/motor assemblies.

Table 1 Comparison of noise emission levels between un-enclosed and acoustically enclosed fan assemblies.

| Fan model | without acoustic enclosure dB(A) | with final design of acoustic enclosure dB(A) |
|---|---|---|
| F1 | 91 | 79 |
| F3 | 98 | 78 |
| F5 | 100 | 83 |
| F6 | 97 | 83 |
| F10 | 94 | 83 |
| F11 | 96 | 85 |
| F12 | 99 | 86 |

For the smaller fan assemblies the acoustic box had very little effect upon the fan assembly performance, whereas the F10, F11 and F12 fans all showed slight loss in performance. The loss however, was less than normally experienced on un-enclosed systems when safety grilles are attached to the fan outlet and hence the gain in reduction of noise emission outweighed the minor performance loss.

It will be noted that the acoustic enclosure design with the modification to the fan orientation was equally effective for reduction in noise emission for all assemblies, but that the desired level had not been achieved for the largest unit.

## 4.7 Check on fan performance

As a check on fan assembly performance a full test was conducted to monitor pressure drop across an F10 fan assembly against volumetric flow. The graph illustrated in Figure 4 shows the effects of exhaust grille and acoustic enclosure compared to the performance of an entirely free running assembly.

The typical fan operating point is indicated on the diagram. It can be seen that the acoustic enclosure offers an improvement of approximately 20% over a fan system using a grille, although the enclosure drops the performance by 20% when compared to a completely free assembly.

## 4.8 Alternative acoustic linings

As a check on the efficiency of the lining material various permutations were tried. Firstly the thickness of lining in the enclosure and additionally in the Header unit were investigated by doubling up the material and testing for noise emission levels. The results obtained are illustrated in Table 2.

Improvements in noise emission levels are achievable as the foam lining is thickened, but care must be taken to ensure that the internal volume lost in the acoustic enclosure must not impinge upon the air flow stream and create further throttling effects.

Table 2 Effect of acoustic lining thickness on noise emission levels for F5 and F12 fan assemblies

| Lining type | F5 assembly dB(A) | F12 assembly dB(A) |
|---|---|---|
| Assembly un-enclosed | 100 | 99 |
| Acoustic enclosure with single layer of foam | 83 | 86 |
| Acoustic enclosure with double layer of foam | 81 | 85 |
| Acoustic enclosure with single layer of foam plus Header unit with single layer lining | 80 | 85 |
| Acoustic enclosure with double layer of foam plus Header unit with single layer lining. | 78 | 84 |

Commercially it is also important to consider the extra costs involved in both labour and material as more material is used. For these reasons it was not considered desirable to investigate beyond double lined enclosures and single lined Header units.

An alternative approach was to alter the lining material and some attention was paid to this aspect. Acoustic lining materials with five times the density of the material previously used produced only a 1 dB(A) reduction in emitted noise and as the cost was proportional to the density of the material it was not considered economically viable. Barrier mats with foam lining for reduction of transmitted vibration to panels were also found to offer little advantage for this application.

## 5 NOISE EMISSION REDUCTION FOR HIGHER TEMPERATURE AIR STREAMS

Some applications require particle extraction from air flow streams where the process raises the temperature of the air flow above the $60^{\circ}$c limit imposed by the electric motor manufacturers to ensure motor efficiency. Clearly for these applications it is not possible to enclose the motor inside the acoustic enclosure. But the alternative solution of mounting the motor external to the enclosed fan system also presents problems, as the motors generate noise from their on-board cooling fans and the enclosure volume around the fan would be too small to ensure that no throttling effect occurred.

For these solutions it was decided to maintain the current attenuator splitter units which make use of perforated surfaced laminates running along the length of typically a 150 cm exhaust duct. To complete the comparison of noise emission an F12 fan assembly was monitored and was found to reduce the noise emission for an un-enclosed system from 99 dB(A) down to 90 dB(A). This reduction is not as good as the acoustic enclosure solution with a single layer of acoustic lining ( where 86 dB(A) was achieved), but offers a reasonable solution for systems requiring noise attenuation for higher temperature flow problems.

## 6    NATURAL FREQUENCIES OF FAN IMPELLER

To derive a fuller understanding of the noise generating mechanism of the fan assemblies investigations were conducted into determining the natural frequencies associated with the structure of the fan impeller system.

The range of fans currently in use are predominantly fabricated from metal sheet which is stamped, folded and then riveted, bolted or welded. Majority use mild steel plate but for fire hazard applications non-ferrous materials are used.

The steel fans exhibit all the vibrational characteristics of structures and inspite of riveted joints they exhibit fairly low damping coefficients. For this reason it was considered important to identify natural frequencies and to relate them to the frequency components in the noise emitted from the fully assembled extractor.

The method used was based upon conventional modal analysis techniques using a dual channel frequency analyser. Excitation of the impeller was effected by both striking the structure with a force instrumented hammer and by exciting with an electromagnetic vibrator driven from a white noise signal source. A typical frequency response spectrum is shown in Fig 5, where peaks in the graph indicate specific natural frequencies identified by that unique test with its specific input and measuring location. Many such tests have to conducted on the structure either by keeping the monitoring position constant and changing the excitation point or vice versa. With the data obtained it is possible to manipulate it in such a way as to be able to identify the phase relationship of all points on the structure at each discrete natural frequency. Hence it is possible to determine the mode shape for each specific frequency and to determine the strength of the mode. The results identified several significant vibrational mode shapes, one of which is shown in Fig 6 where the 2D (two diameters across the face of the fan) is illustrated.

The significance of such an analysis is that it identifies frequencies easily stimulated in the fan and highlights possible weaknesses in design which are likely to be vulnerable to fatigue failure. It allows for the possibility to alter design philosophy by giving an insight into how natural frequencies can be shifted away from critical areas such as the blade pass frequency. Hence decisions such as stiffening or weakening a particular structural element or even adjusting the number of blades can be taken in the knowledge of what is likely to happen to the structures natural frequencies.

This analysis process is continuing and will be used to incorporate a full understanding of the part played in noise emission by the fan casing, motor mounts and structure of the full extractor system

## 7    FAN DESIGN

Alongside the noise and vibration analysis, a second programme is looking into the fan design parameters with a view to identifying optimum designs to ensure minimum noise emission. Fan blade geometries, clearances, scroll geometries, blade configuration and exhaust ports are all currently under investigation.

## 8    CONCLUSION

Reduction in the noise emitted from equipment such as dust extraction machinery is a complex problem which has to be studied using a wide variety of modern techniques in order to effect the maximum gains. The increasing public awareness of the hazards of noise in the environment are forcing designers to seek progressively lower levels of emission. It is necessary therefore to approach the problem from both aerodynamic and structural considerations.

ACKNOWLEDGEMENTS

The authors wish to thank DCE Group Ltd for allowing them to report investigations on their products, the Teaching Company Directorate for supporting the programme of work and Loughborough University of Technology for assisting with technical expertise and research equipment.

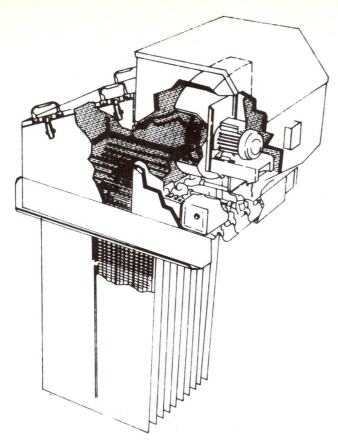

Fig 1    Sectional view of a dust extraction system

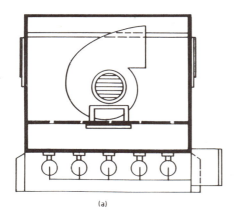

(a)

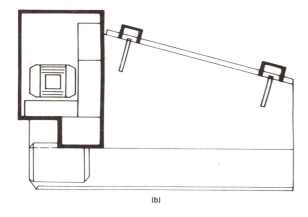

(b)

Fig 2    Acoustic enclosure boundaries
(a) plan view and (b) side view

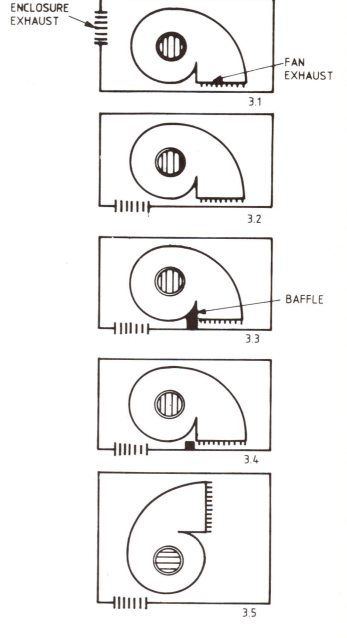

ENCLOSURE EXHAUST

FAN EXHAUST

3.1

3.2

BAFFLE

3.3

3.4

3.5

Fig 3    Acoustic enclosure configurations
Fan exhaust — enclosure exhaust
3.1  Vertical down — Horizontal (side panel)
3.2  Vertical down — vertical (bottom panel)
3.3  As for 3.2 plus full internal baffle
3.4  As for 3.2 plus half internal baffle
3.5  Horizontal (right) — vertical (bottom panel)

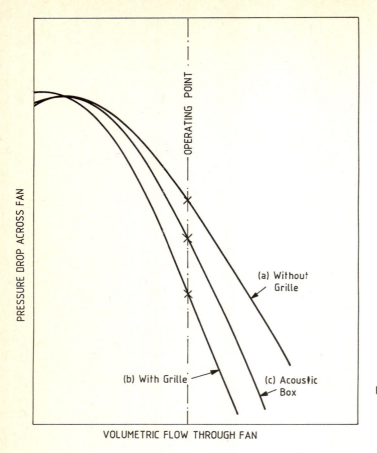

Fig 4    Performance characteristics for F10 fan assembly
(a) free running system
(b) un-enclosed fan with safety grille on exhaust
(c) assembly with acoustic enclosure

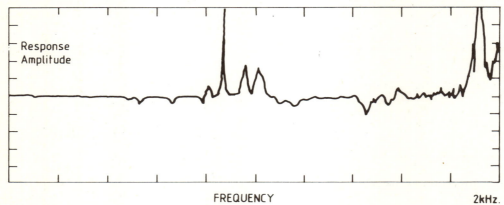

Fig 5    Typical frequency response spectrum for fan impeller

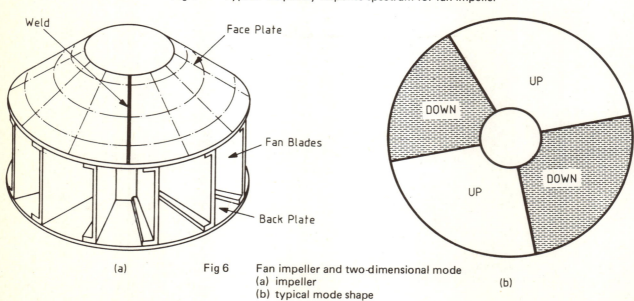

Fig 6    Fan impeller and two-dimensional mode
(a) impeller
(b) typical mode shape

# C401/031

# Noise breakout through the walls of fan system ducting

A CUMMINGS, PhD, DEng, CEng, FIOA
Department of Engineering, Design and Manufacture, University of Hull

The mechanism of the transmission of internally propagated acoustic noise through the walls of sheet metal ductwork having various cross-sectional geometries is described. Measured and predicted data for ducts having rectangular, flat-oval and circular cross-sections are compared. Some simplified wall transmission loss (TL) formulae are presented.

## 1.    Introduction

Noise radiation from the walls of sheet metal air-moving ductwork in buildings is an ever-present problem that faces the heating and ventilating engineer who is concerned with acoustical design. Aerodynamic sound, generated either by fans or by turbulence at bends or other discontinuities, propagates along the ductwork and causes the duct walls to vibrate and to radiate noise into the surrounding space. This phenomenon is known as "breakout". In typical duct installations, breakout is a problem principally at low frequencies and manifests itself as a "rumble". This characteristic adds to the difficulty, since low-frequency noise is generally not easily attenuated.

A related problem is that of noise transmission into ducts (leading to subsequent radiation from the walls or from terminal units); this is termed "breakin". The two phenomena go hand in hand and it is in fact possible, by the use of the acoustic principle of reciprocity, to predict the breakin characteristics of a given duct once the breakout characteristics are known, as Vér [1] has shown.

A means of calculating breakout is therefore necessary in noise control procedures and, of course, the physical mechanism of breakout must be properly understood.

Until the late 1970s, very little attention had been devoted to the understanding of breakout from sheet metal ducts, most of the previous work having been concerned with sound transmission through perfectly circular cylindrical shells or pipes; these problems, as it turns out, have little relevance in the context of heating, ventilating and air-conditioning (HVAC). A simplistic formula reported by Allen [2] was widely used for estimating breakout, but is based on the assumption that many acoustic modes propagate in the duct, and this condition does not hold for typical ductwork in the low frequency region where breakout is usually a problem. There is no doubt that Allen's formula is quite inadequate for predicting low frequency noise breakout from ducts, and that a much more detailed approach is required.

The author and his co-workers (see, for example, references [3-7]) have examined the mechanism of noise breakout from sheet metal ducts having rectangular, circular, distorted circular and flat-oval cross-sectional geometries in some detail and have devised methods for predicting the radiated noise in each case. A variety of methods of prediction is described in references [3-7] and in other related publications, and in certain cases it has been possible to express these results in simple formulae; in other cases, more complicated procedures or rather lengthy computation is necessary. Comparison between predicted and measured breakout levels was made in this work, and the agreement between experiment and theory was found to be generally satisfactory. Experimental studies of breakout have been conducted by Guthrie [8] and by Nelson and Burnett [9]. Very little other work related to noise breakout from HVAC ductwork appears to have been carried out. Almgren [10] gave a good general account of noise breakout from ductwork, though this is now a little out of date in view of subsequent research.

In the present paper, the physical processes underlying noise breakout from ductwork are described, though lengthy mathematical derivations and formulae are deliberately avoided. Instead, emphasis is placed on the qualitative aspects of the problem, and on the results of the various analyses and tests.

## 2.    Some General Aspects of Noise Breakout Through Duct Walls

Consider a straight duct of arbitrary but uniform cross-sectional geometry as shown in Figure 1. The duct walls are rigid except for a finite section that is flexible. The structural boundary conditions at the junction between flexible and rigid sections need not be defined for the purposes of the present discussion. Imagine that sound waves (of an unspecified nature, but shown for simplicity as plane harmonic waves in Figure 1) travel inside the duct from left to right. Though mean air flow would normally exist within the duct, it can be ignored because of the relatively low flow speeds in HVAC ducts. In what follows, it will be assumed that air flow effects are negligible. Moreover, vibrational excitation of the duct walls from fluctuating fluid stresses at the wall in a turbulent boundary layer will also be neglected here. Though it is not immediately obvious, this means of excitation is of secondary importance to acoustic excitation in practical cases.

Within section I of the duct (the left-hand rigid section), the duct walls cannot vibrate and therefore no external radiation occurs. But as soon as the sound waves encounter the flexible section, the acoustic pressure changes cause the walls to vibrate. The nature of this vibration depends on a number of factors: (i) the character of the internal sound field, including the transverse acoustic pressure distribution and the frequency content, (ii) the speed of sound in, and the densities of, the fluids inside and outside the duct (in HVAC applications, both fluids would be air having roughly the same properties), (iii) the thickness, density, Young's modulus and Poisson's ratio of the duct walls, and (iv) the cross-sectional geometry of the duct. Other factors that are involved to a lesser extent are: (v) the length of the flexible section of duct, and (vi) the boundary conditions at either end of the flexible section.

## 2.1 Duct Wall Vibration

Quite clearly the wall vibration will, in general, be of a highly complex nature. Fortunately, the main features, as far as noise radiation is concerned, may be embodied in a relatively simple model. This involves the idea of individual acoustic "modes" (or propagating pressure patterns, each of which is a solution to the governing acoustic wave equation) inside the flexible portion of duct, forcing structural wave motion in the walls travelling at the same phase speed as the acoustic modes. Strictly speaking, these "structural/acoustic modes" should be considered to be coupled: that is, the interaction between the internal sound field, the wall vibration and the external sound field should be properly taken into account via the appropriate boundary conditions. This coupling complicates the issue, however, and, except close to transverse structural resonance frequencies of the duct walls, it may be neglected.

What remains is an uncoupled vibrational motion of the walls, forced by the internal sound field, the external acoustic radiation load on the walls being ignored. Provided the internal sound field can be specified in some suitable way, and the equations of motion for the walls can be solved with a forcing function that is representative of the internal sound field (an eigenfunction expansion is a particularly convenient way of writing this forcing function), then the wall vibrational pattern can be found. In this process, the structural boundary conditions at the ends of the flexible portion of duct are required. Again, the actual situation may be simplified: one may neglect the presence of these axial boundaries and assume that structural

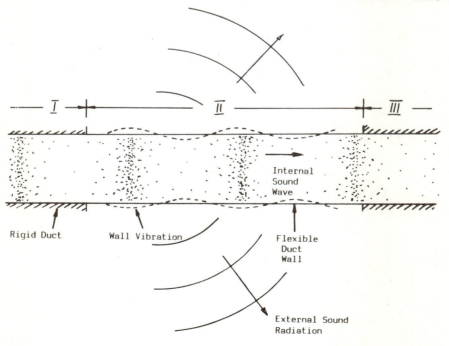

Fig 1    Acoustic breakout from a duct with flexible walls

waves, travelling only in the direction of propagation of the internal sound waves, are present in the duct walls, and that these are unaffected by the presence of the boundaries (this implies that no wave reflection occurs at the boundaries). Each internal acoustic mode or eigenfunction travels at a characteristic phase speed and is considered, in the present treatment, to give rise to a component of the wall vibration pattern that travels at the same wave speed. If more than one acoustic mode travels inside the duct, then the wall vibration is made up of the sum of the vibrational patterns produced by the various acoustic modes, since both the sound field and the wall vibration are taken to be linear. Because structural wave reflections at the axial boundaries of the flexible section of duct are ignored, the above assumptions give rise to a "forced wave" model, involving only structural motion made up of components each having a direction of travel and wave speed that are the same as the corresponding modal components of the internal forcing acoustic pressure field. "Free wave" effects, involving structural wave reflections, are not included. The generally good agreement between forced wave theoretical models and measured data serves to justify the use of forced wave models for the structural response of the walls.

The nature of the wall vibration pattern will clearly be dependent upon factors (i) - (vi) above. The modal and spectral contents of the internal sound field will have a considerable effect. The frequency distribution would generally be of greater significance, since the excitation frequency of the wall motion would determine structural resonance effects, which are usually quite sensitive to frequency. The speed of sound in, and the density of, the fluid within and external to the duct would usually be those of air at roughly STP, so factor (ii) would not normally be of any significance. The mechanical properties and thickness of the duct wall, (iii), will in themselves have a substantial effect on the sound transmission characteristics of the duct, though in practical cases the wall material would usually be galvanised steel having a limited range of thicknesses. The cross-sectional geometry, (iv), of the duct is of overriding importance, and is the main factor that determines the TL. The length of the flexible section of duct, (v), has a fairly minor effect on the TL, as do the axial boundary conditions (vi).

## 2.2 Radiation from the Duct Walls

Space does not permit a lengthy discussion here of sound radiation from duct walls to the exterior region. Brown and Rennison [11] described a simple line source model for sound radiation from a pipe wall, and gave expressions for radiated sound power. Cummings [4] has extended this work and discussed various line source models in detail, and Astley and Cummings [12] have made comparisons between a line source model and the results of a finite element analysis of round radiation from a duct. It was shown, in this latter piece of work, that a simple line source model, modified to account for higher order mode sound radiation, gave acceptably good predictions. Cummings, Chang and Astley [6] described a theoretical model for sound radiation from circular ducts of finite length, again making comparison with finite element computations.

A simple picture of low frequency sound radiation from the duct wall is as follows. Assume that only the plane acoustic mode propagates inside the duct, at the speed of sound. Then (according to our forced wave model) the duct wall will carry structural waves travelling at the acoustic speed, from one end of the flexible portion of duct to the other. A vibrational volume velocity per unit length of duct wall may be defined (this will be equal to the vibrational velocity amplitude, integrated around the duct's perimeter), and this may be put equal to the volume velocity per unit length of peristaltic waves travelling along a finite length line source in free space, the length of which is equal to the length of the flexible section of the duct. If the radiating length of duct is infinite, one can show that no sound power at all emanates from the duct walls. In reality, however, acoustic scattering occurs at the ends of the radiating section, resulting in the radiation of energy. A "radiation efficiency", $\sigma$, may be defined as the ratio between the sound power, per unit length of duct, radiated from the finite length line source and the sound power per unit length radiated from a line source of infinite length, carrying structural waves of supersonic wave speed, if all other factors remain the same between the two situations. In many cases of practical interest, that is for ducts that are fairly long compared to the acoustic wavelength, and for structural wave speeds close to the acoustic speed, $\sigma$ is approximately 0.5. This is a good rule of thumb in the absence of any detailed information. The above argument applies to sound radiation that emanates uniformly from the duct in the azimuthal direction. "Higher order" acoustic radiation, with more complex patterns, can also carry energy (see reference [6]), though the radiation efficiency would normally be small at low frequencies.

The reader must wonder whether this rather idealised picture of sound radiation from a line source is actually realistic in the context of HVAC systems. Experience, and comparisons between between experimental and theoretical data (see reference [12]), indicate that it is; the length of a given section of exposed ductwork within an internal building space should be used as the length of the equivalent line source in these calculations.

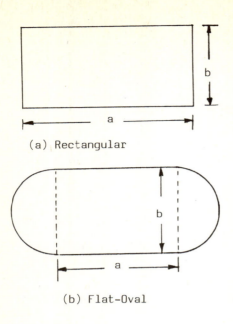

(a) Rectangular

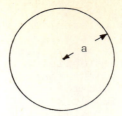

(c) Circular

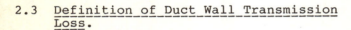

(b) Flat-Oval

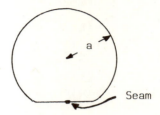

(d) Distorted Circular
("Long Seam")

Fig 2    Cross-sectional shapes of ducts

### 2.3  Definition of Duct Wall Transmission Loss.

It is difficult to choose an altogether satisfactory TL definition for breakout.  In the case of sound transmission through a flat building partition, the TL may straightforwardly be expressed as a logarithmic ratio between either incident and transmitted sound powers or intensities; since the areas over which the sound is incident upon, and transmitted by, the partition are the same, the two definitions are identical.  Where sound is radiated by a duct wall, however, the sound energy is incident over the cross-sectional area of the duct but radiates over the exposed surface area of duct.  Since these areas are not the same, definitions of TL in terms of sound power or intensity will differ.

We should like to have a TL that (i) is not a function of duct length, but rather of wall material, cross-sectional geometry and frequency, and (ii) tends to zero as the duct wall becomes transparent to sound.  These two requirements turn out to be mutually exclusive in the case of breakout.  Requirement (i) would normally take precedence, since it would be preferable not to have the TL dependent to any great extent upon the length of the duct.

The TL definition adopted here is as follows.  Let L be the radiating length of the duct, P the perimeter and A the cross-section area, $W_0$ the sound power entering the duct and $W_r$ the total sound power radiated from the walls.  We then define

$$TL = 10 \log \left[ \frac{W_0}{A} \cdot \frac{PL}{W_r} \right] \qquad dB, \qquad (1)$$

this being the dB ratio between the average sound intensity (sound power per unit cross-sectional area) entering the duct and the average sound intensity radiated from the duct.  Because of the nature of sound radiation from the duct walls, the TL will be rather weakly dependent upon L, but this dependence may be ignored expect at very low frequencies.

### 3.  The Mechanisms of Noise Breakout through Duct Walls of Various Geometries.

In this section, we discuss the mechanism of noise breakout in more detail and, in particular, highlight the effects of cross-sectional geometry of the duct on the structural response and thus on the TL.  Figure 2 shows the cross-sectional shapes of the commonest types of duct; these will encompass virtually all practical duct installations.

### 3.1  Rectangular Ducts

Consider a duct of rectangular section, as depicted in Figure 2(a).  At low frequencies, the internal sound pressure perturbation will be uniform over the cross-section and the waves will travel at approximately the speed of sound in air.  As previously mentioned, the structural waves in the wall are also assumed to travel at this speed.

To examine the wall motion in detail, we may write the equation of motion for each duct wall, including a forcing term representing the acoustic pressure variation inside the duct, and thus obtain "forced wave" solutions to these equations subject to appropriate structural boundary conditions at the duct corners.  This

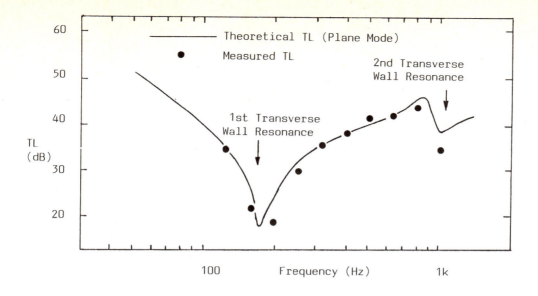

Fig 3    Theoretical and measured TL of square section mild steel duct,
$a = b = 203$ mm, wall thickness = 1.22 mm

process was described by Cummings [3,14]. The boundary conditions chosen are (i) zero normal displacement at each corner, (ii) continuity of bending movement about the corner between adjacent walls, (iii) that the duct corners should remain right-angled as they rotate. The solution to the equations of motion reduces to the simultaneous solution of eight linear equations, which may readily be accomplished by standard methods. Details of the wall motion and any resonance effects are obtained from this process. In most rectangular ducts, there is a seam along one or more corners. This does not have any significant effect on the wall TL in general, though it can alter the boundary conditions at the corner, and hence change the transverse resonance frequencies of the duct walls; Astley and Cummings [12] discussed these effects.

The characteristic "rumble" that is so often noticeable as a subjective feature of breakout noise may be associated partly with low frequency transverse resonances of the duct walls. This is particularly the case with square section ducts, as illustrated by Cummings [3]. These resonances are present over a wide range of frequencies, but are progressively more heavily damped as the frequency rises, and so are only of significance at low frequencies, usually below 200 Hz. The "wave theory", just described for the analysis of the wall motion, yields details of these resonance effects.

Apart from the low-frequency structural resonances, there is an overall slope to the TL curve (versus frequency) of +3dB per doubling of frequency. Especially with the larger sizes of duct, which have very low fundamental structural resonance frequencies, this is the principal feature of TL curves for rectangular ducts. If resonance effects

are not considered to be important, then simplified physical models may be adopted to describe the duct wall response, based on a limp mass behaviour of the walls. Naturally - since the spring-like characteristics of the walls are ignored - resonance effects are absent from such models, but this may be of little importance in engineering calculations, particularly for ducts of fairly large dimensions. Even though the duct wall vibration is found on a simplified basis, the sound radiation from the duct walls must still be determined from the line source model or some equivalent. As far as the internal sound propagation is concerned, the duct walls may still be assumed rigid. Cummings [5] gives "approximate asymptotic solutions" for the TL, based on the above assumption. These take particularly simple forms if $\sigma$ is assumed to be 0.5 and if the duct walls are not too transparent to sound. Different expressions apply in the cases where (i) only the fundamental acoustic mode propagates within the duct, (ii) many (about 10 or more) modes propagate. In case (i), we have

$$TL = 10 \log [4\omega q^2/\rho^2 c (a + b)] \qquad (2a)$$

and in case (ii),

$$TL = 10 \log (\omega^2 q^2/7.5 \rho^2 c^2); \qquad (2b)$$

here, $\rho$ is the air density, $c$ the sound speed, $q$ the mass per unit area of the duct walls, $\omega$ the radian frequency ($2\pi$ x frequency) and $a,b$ are the transverse dimensions of the duct (see Figure 2(a)). Equation (2b) results from a "statistical" treatment, where the collective properties of the modes at a particular frequency are considered, rather than those of the individual modes themselves. Clearly, the TL will tend toward a value of 10 log (PL/A) at very low frequencies, where $W_r \rightarrow W_o$, and this therefore imposes a lower limit on the TL: equation (2a) will hold

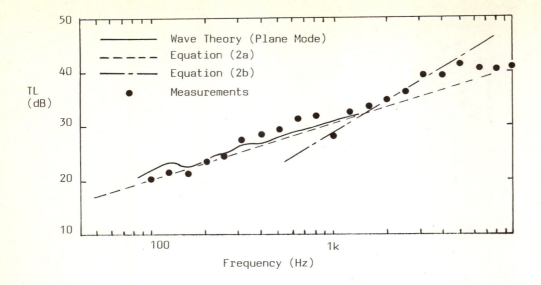

Fig 4    Theoretical and measured TL of rectangular section galvanized
steel duct, $a$ = 762 mm, $b$ = 356 mm, wall thickness = 0.64 mm

if the predicted TL is greater than 10 log (PL/A), otherwise this limiting value will apply. Equation (2b) should be valid for frequency $\geq 613/\sqrt{ab}$ (a,b in metres), and equation (2a) should be used at frequencies below this limit.

Figure 3 shows measured TL data taken on a 203mm square mild steel duct, together with predictions from the wave theory. Agreement between experiment and theory is observed to be good, the first transverse structural resonance (manifested by the dip in TL at 170 Hz) being predicted accurately. In this particular duct, the axial seam was butt-welded and beaten flat, so the assumed boundary conditions would have been satisfied. It may be noted that when tape-recorded fan noise was fed into one end of the duct from a loudspeaker, the radiated noise had the subjective quality of narrow-band noise centred around 170Hz. This duct is, however, not really representative of HVAC ducting, being at the small end of the range of transverse dimensions, having rather thicker walls than usual and also being constructed in an atypical way.

We may take the data in Figure 4 as being rather more representative of rectangular HVAC ducts. The measured data were taken on a duct of dimensions 762mm x 356mm. Equation (2a) predicts the TL quite well below about 1.5 kHz. Above this frequency, equation (2b) predicts a 6dB per octave slope, and this is observed in the measurements up to about 6.3 kHz. The wave theory is broadly in agreement with equation (2a), and (as expected) shows undulations caused by damped resonances (a structural loss factor is included in these predictions, and it usually has to be given a very large value, of the order of 0.1-0.2, to give

quantitatively good agreement between predictions and measurements near wall resonance frequencies). This duct had seams along two opposite corners, causing rather less than rigid behaviour. Nonetheless, the wave theory gives good predictions of the TL in the region where it should be valid.

Equations (2a) and (2b) are generally satisfactory for predicting the TL of rectangular ducts and tend to give results that are, if anything, slightly conservative for larger sized ducts.

## 3.2 Flat-Oval Ducts

Figure 2(b) shows the geometry of flat-oval ducts. Analysis of the wall motion is rather more complicated than it is in the case of rectangular ducts, since three coupled equations of motion are required to describe the response of the circular walls.

Two approaches to the problem have been utilised by the author. In the first, the sound field in the duct is assumed to consist of "hard-wall" modes, the structures of which are found by a numerical solution to the acoustic wave equation over the duct cross-section, except in the trivial case of the plane mode. An appropriate summation of modes is assumed (in the simplest case, this would consist of the plane mode alone), and the wall response is found for each mode from a numerical solution to the equation of motion of a uniform cylindrical shell of flat-oval cross-sectional shape. The sound power radiated from the duct wall by the summation of model structural responses is found; the internal sound power is determined from the assumed combination of acoustic modes within the duct, and the TL is then calculated.

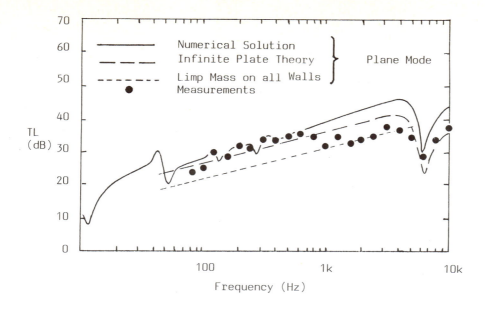

Fig 5    Theoretical and measured TL of flat-oval section galvanized steel
duct, $a$ = 522 mm, $b$ = 254 mm, wall thickness = 0.64 mm

The second method has been implemented only for the plane internal mode, and involves finding the duct wall response separately on the flat and the curved walls, from impedance expressions appropriate to <u>infinite</u> flat and curved plates (the idea of a circular cylindrically curved plate of infinite extent in the azimuthal direction is hard to grasp, but the assumption really signifies the absence of azimuthal boundary effects). A composite wall response is thus found, and the radiated sound power determined on the basis of this.

This second method is relatively simple and can easily be applied <u>via</u> the use of a pocket programmable calculator. The first method, on the other hand, requires considerable computation, particularly where several acoustic modes within the duct are considered. An obvious development of the aforementioned analyses would be a statistical approach similar to that applied in the case of rectangular ducts. This has not so far been carried out; the flat-oval geometry presents considerable additional complications, over and above those encountered in the case of rectangular ducts. In principle, however, a statistical TL theory is possible.

Flat-oval ducts would appear to be a hybrid of rectangular and circular ducts, since they have both flat and curved walls. Indeed, a simplistic view of the physics of the wall vibration might lead one to suppose that features common to both rectangular and circular ducts might appear in the case of flat-oval ducts. At low frequencies, rectangular ducts exhibit a kind of "oil-can" motion, where the wider walls deform most in the centre and least near the corners; the narrower walls

deform less and thus make a smaller contribution to the radiated sound energy. At higher frequencies, the wall motion becomes more complicated, but well above the fundamental wall resonance, the response is essentially mass-controlled, and the TL curve has a slope of 3dB per octave. The walls of circular ducts present an extremely high impedance to the internal sound wave at low frequencies and the TL is therefore high, though it falls as the frequency increases. A minimum in the TL is reached when the walls exhibit a "ring resonance", that is, a uniform resonant motion analogous to the fundamental resonance of a ring made by cutting a narrow transverse slice from the duct wall. A "wave coincidence" effect also occurs in ducts of all geometries, at the frequency where the free axial bending-wave speed in the walls is equal to the sound speed; a maximum wall response and TL minimum result at the coincidence frequency. In typical HVAC ducts, however, the coincidence effect occurs above 10kHz and is thus of little practical interest.

It might therefore be expected that flat-oval ducts would show a 3dB per octave slope, at low frequencies, in the TL curve, where the flat walls behave like those of rectangular ducts and the curved walls vibrate relatively little. Equally, it would seem possible that the curved walls would bring about a ring-type resonance at a higher frequency. Simplistic notions of this sort are often erroneous, however, and one must look to measured or theoretical data for evidence of these effects.

Figure 5 shows TL measurements on a flat-oval duct with a = 522mm, b= 254mm (see Figure 2(b)), together with theoretical curves derived from a finite difference solution to the equations of motion of the walls, the "infinite plate" model previously described, and an assumption of limp mass behaviour on all walls. The theoretical curves are all for the plane acoustic mode. Agreement between measurements and the numerical solution is good up to 800 Hz. The infinite plate model also gives good predictions of the TL. The limp mass impedance model gives an underprediction of the TL at low frequencies. Two main features are evident in the measured data and in the first two theoretical curves: first, a 3dB per octave slope in the TL at low frequencies, and secondly, a ring resonance at the expected frequency. Apparently the duct walls do, indeed, act as a hybrid of rectangular and circular geometries. Predicted wall displacement patterns reveal that an "oil-can" motion of the flat walls exists at low frequencies, with the curved walls acting like springs. As the frequency rises, more complex wave patterns develop in the walls and close to the ring frequency, the circular walls display a more or less uniform pulsatory motion over most of their extent, with the vibrational amplitude of the flat walls being much smaller.

A further point to be noted in Figure 5 is that the flat-oval duct has a higher TL than that of the equivalent rectangular duct (the limp mass theoretical curve) at low frequencies. This is because the circular walls do not contribute significantly to the sound radiation. Clearly, flat-oval ducts have an advantage, in TL, over rectangular ducts. The smaller the ratio a/b, the greater will be the difference between the TL of the flat-oval duct and the equivalent rectangular duct.

### 3.3  Circular and Distorted Circular Ducts

These are shown in Figures 2(c) and 2(d). Ducts with a perfectly circular cross-section have, as mentioned in Section 3.2, a very high TL at low frequencies because the wall motion is uniform (for plane internal sound waves) and involves tangential compression and tension of the walls rather than bending. But most practical "circular" HVAC ducts are not actually circular at all, and have at least some degree of distortion from circularity. There are two common kinds of circular duct: "spiral-wound" ducts, wound from a single strip of galvanised steel, and "long-seam" ducts, with a single axial seam. The latter type is illustrated in Figure 2(d); there is, typically, a flattened region along either side of the seam.

The effect of wall distortion is to give rise to what has been termed "mode coupling". In this, an internal acoustic mode, with its characteristic sound pressure distribution, will excite not only the structural mode with the corresponding displacement pattern, but also other structural modes, because the

wall vibration enables the forcing pressure distribution to couple to a series of structural modes, the precise nature of this coupling depending on the distortion profile. For example, consider plane sound waves propagating within a duct at a low frequency. If the duct cross-section is, say, slightly elliptical, it is very easy to visualise an acoustic pressure maximum causing the duct to tend more toward a circular shape, with the reverse happening at a pressure minimum. (This effect is well-known: a flattened hosepipe will become circular when pressurised, and flatter if a negative pressure is applied.) In this case, the plane sound wave will excite an "ovalling mode" in the walls with four displacement modes around the duct's perimeter.

Mode coupling effects caused by the plane acoustic mode involve very much greater wall displacements than those resulting from the uniform wall motion associated with a perfectly circular duct. The consequence of this is that the higher structural modes will themselves radiate significant acoustic energy. Though the radiation efficiency of these modes is considerably less than that of the uniformly pulsating mode, they still radiate far more sound energy, in total, than that from the uniform structural mode. The difference is often so large that the sound radiation by the uniform mode is quite negligible compared to that from the higher modes. A lowering of TL by up to 60dB at low frequencies, brought about by mode coupling, is common! This is clearly a very great effect, of such magnitude that the TL mechanism expected on the basis of an ideally circular duct is actually quite irrelevant, at low frequencies, in the practical case of a distorted sheet metal duct. As the frequency rises, mode coupling effects become progressively less important, since the ideal duct TL falls with increasing frequency. Even in the presence of mode coupling effects, the TL of a typical distorted circular duct is still generally much higher, at low frequencies, than that of the equivalent rectangular or flat-oval duct.

Prediction of the TL of distorted circular ducts is difficult at best. It is complicated by the fact that a detailed knowledge of the variation in local radius of curvature of the duct wall, around the perimeter, is necessary, even for ducts with uniform distortion along their length. Having measured the radius of curvature (which is in itself a tedious procedure), one needs to find the duct wall response from the solution to the equations of motion of a non-circular cylindrical shell with a prescribed forcing pressure distribution. The author has investigated two methods. The first involves writing the equations of motion in approximate form (rejecting terms that are likely to be small), expressing the wall distortion as a Fourier series and seeking solutions to the wall displacement

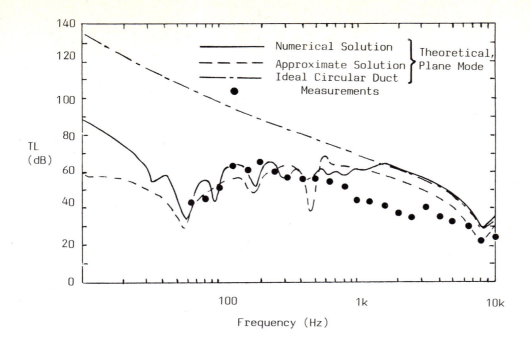

Fig 6    Theoretical and measured TL of distorted circular section long-seam
galvanized steel duct, *a* = 102 mm, wall thickness = 0.51 mm

in the form of Fourier series. This may
be termed the "approximate" method. The
second method involves the solution to the
equations of motion by the use of finite
difference approximations.

Figure 6 shows measured TL data on a
long-seam circular duct of radius 102mm,
together with distorted-duct predictions
based on the numerical solution and the
approximate theory, and an ideal duct
theoretical TL curve. It is immediately
apparent that the ideal duct theory fails
totally to predict the TL. At 63Hz, the
TL is overpredicted by about 60 dB. At
higher frequencies, the discrepancy is
less, but is still substantial. Both the
approximate theory and the numerical
solution are at least qualitatively in
agreement with the measurements up to
about 630 Hz. The numerical solution is
in generally better agreement, and at most
frequencies within this range is fairly
close to the experimental data. It may be
noted that the measured TL is quite high,
at low frequencies, as compared to that of
a rectangular or flat-oval duct of similar
dimensions. Between 63Hz and 500Hz, the
TL lies between 44dB and 66dB. As
expected, all theoretical curves and the
experimental data show a ring resonance at
8 kHz.

It would appear that the mode-
coupling mechanism adequately explains the
shortfall of the measured TL at low
frequencies as compared to the ideal duct
TL. On the other hand, implementation of
either of the theoretical analyses
described here is extremely tedious and
certainly not within the scope of ordinary
engineering design calculations. It is
possible that a standard form of wall

distortion could be assumed for long-seam
ducts, and a series of design charts
produced, but this would not be applicable
to spiral-wound ducts, which have no
readily identifiable mode of distortion.
An extensive series of tests on long-seam
and circular ducts [13] indicated that the
low-frequency TL can vary considerably
between individual ducts, depending on the
extent of wall distortion. Long-seam
ducts tend, in general, to have a
considerably lower TL, at low frequencies,
than equivalent spiral-wound ducts,
because of their inherently greater degree
of distortion.

## 4.    The Reduction of Noise Breakout from Duct Walls

In the acoustical design of HVAC
systems, considerations of breakout will
often result in a requirement for noise
reduction. In existing systems where
duct-radiated noise is a problem, curative
measures may also be required.
Accordingly, we need to consider methods
of attenuating breakout noise from ducts.

The preceding discussion has revealed
that, in the critical low-frequency
region, "similar" ducts of the three
geometries discussed have increasing low
frequency TL (at the same frequency) in
the order: rectangular, flat-oval and
circular. Clearly, from a design point of
view, circular ducts will be preferable to
the other two kinds, and in general
spiral-wound ducts would be less prone to
mode-coupling effects than long-seam
ducts. In some situations, considerations
of space within a ceiling void would
preclude the use of circular ducts, and
here, flat-oval ducts should be used, with
as low a ratio a/b as possible.

Rectangular ducts, having the poorest TL characteristics, should be avoided where noise problems are likely to occur.

The choice of geometry is, of course, only one means of preventing excessive levels of noise breakout, but it is the most economical approach. Supposing rectangular ductwork (for example) must, for one reason or another, be used, then another means of increasing the TL would be to increase the mass per unit area of the walls. This is expensive and imposes increased structural loads on the building. Doubling the mass per unit area would increase the TL by about 6dB.

The application of external lagging to duct walls is a method of noise control that is sometimes used to cure existing noise problems; it is not a very satisfactory means of prevention by design. The lagging consists, typically, of a layer of porous sound-absorbing material such as mineral wool, with an outer impervious layer of material such as plaster, plastic sheet or a metal casing. Cummings [15] has reported a method of predicting the acoustic insertion loss of lagging on the walls of rectangular ducts. This principle has been extended, in approximate form, to apply to flat-oval and circular ducts [13]. It transpires that lagging can bring about a reduced TL in a narrow frequency band, typically at frequencies of 50-300 Hz, because of a damped resonance effect. Only above the resonance frequency is any really useful attenuation obtained. The resonance frequency may be lowered - and the useful frequency range of the lagging extended - by either (or both) increasing the mass per unit area of the lagging or increasing the thickness of the absorbent layer. Around the resonance frequency, however, it is quite possible for the radiated noise level to be increased, and the problem exacerbated, by the application of lagging. This method of noise control cannot be regarded as being generally very satisfactory at low frequencies, and most certainly must be used with caution.

## 5.    Discussion and Conclusions

The essential physical mechanisms of noise breakout from ducts may be understood on the basis of uncoupled, forced-wave models. Predictions made on this basis are broadly in good agreement with experimental data.

It has been shown that the main factor determining the duct wall TL is the cross-sectional geometry of the duct. This is the reason for the relatively high TL of circular section ducts, and the low TL of rectangular ducts, at low frequencies. Moreover, flat-oval ducts, rather interestingly, display features of both rectangular and circular ducts.

The control of breakout noise is best effected by the judicious choice of cross-sectional geometry rather than by the use of "add-on" treatments such as lagging.

One phenomenon that has not been mentioned so far in this paper, but which appears to be of some practical importance, is the limiting effect of breakout and breakin on the attenuation, in situ, afforded by internal sound-absorbing lining in ducts. Suppose a flexible sheet metal duct passes through a reverberant building space and is acoustically lined, to reduce the transmission of "system noise" (fan noise, et cetera) along the ductwork. At the upstream end of the lining, the internal sound level will be relatively high, and sound energy will both travel along the duct (being absorbed in the lining) and be radiated out through the duct walls. The breakout sound energy will set up a reverberant round field in the surrounding region, and some energy will re-enter the duct. This will be of a relatively low level and will not materially affect the internal sound level. Farther downstream, however, the internally transmitted sound level will have been much reduced by the lining and the breakin sound energy (which will be of much the same magnitude at all points along the duct) will tend to dominate the internal sound field. A "levelling off" of the internal sound pressure level inside the duct will thus occur, as the distance downstream increases, rather than the desired linear decrease in sound level with distance. The effectiveness of the lining will be inhibited by the breakout/breakin energy path, which acts as "flanking" transmission, parallel to the direct energy path. Vér [16] presents experimental data which graphically illustrate this effect. At all events, the phenomenon is of considerable significance in the context of lined duct attenuation, and is currently being investigated by the author.

## References

1.    I L Vér June 1983 Bolt, Beranek and Newman Inc Report No. 5116 (Final contract report on ASHRAE TRP-319). Prediction of sound transmission through duct walls; breakout and pickup.

2.    C H Allen 1960 Noise Reduction (ed L L Beranek), New York: McGraw Hill. See Chapter 21.

3.    A Cummings 1978 Journal of Sound and Vibration 61, 327-345. Low frequency acoustic transmission through the walls of rectangular ducts.

4.    A Cummings 1980 Journal of Sound and Vibration 71, 201-226. Low frequency acoustic radiation from duct walls.

5. A Cummings 1983 _Journal of Sound and Vibration_ 90, 211-227  Approximate asymptotic solutions for acoustic through the walls of rectangular ducts.

6. A Cummings, I-J Chang and R J Astley 1984 _Journal of Sound and Vibration_ 97, 261-286.  Sound transmission at low frequencies through the walls of distorted circular ducts.

7. A Cummings and I-J Chang 1986 _Journal of Sound and Vibration_ 106, 17-33. Noise breakout from flat-oval ducts.

8. A Guthrie 1979 _MSc Dissertation, Polytechnic of the South Bank_. Low frequency acoustic transmission through the walls of various types of ducts.

9. P M Nelson and R Burnett 1981 _Sound Attenuators Ltd Report No. TRC 107_. Laboratory measurements of breakout and break-in of sound through walls of rectangular ducts.

10. M Almgren 1982 _Chalmers University of Technology Department of Building Acoustics Report F82-03_  Prediction of sound pressure level outside a closed ventilation duct - a literature survey.

11. G L Brown and D C Rennison 1974 _Proceedings of the Noise, Shock and Vibration Conference, Monash University, Melbourne_, 416-425. Sound radiation from pipes excited by plane acoustic waves.

12. R J Astley and A Cummings 1984 _Journal of Sound and Vibration_ 92, 387-409.  A finite element scheme for acoustic transmission through the walls of rectangular ducts: comparison with experiment.

13. A Cummings 1983 _University of Missouri-Rolla Department of Mechanical and Aerospace Engineering Final Contract Report on ASHRAE RP 318_  Acoustic noise transmission through the walls of air conditioning ducts.

14. A Cummings 1979 _Journal of Sound and Vibration_ 63, 463-465.  Low frequency sound transmission through the walls of rectangular ducts: further comments.

15. A Cummings 1979 _Journal of Sound and Vibration_ 67, 187-201.  The effects of external lagging on low frequency sound transmission through the walls of rectangular ducts.

16. I L Vér 1978 _ASHRAE Transactions_ 84, 122-149.  A review of the attenuation of sound in straight lined and unlined ductwork of rectangular cross-section.

# C401/013

# Three-dimensional viscous computations of complex flows in ducting systems

**A TOURLIDAKIS**, Dipl-Eng and **R L ELDER**, BSc, PhD
School of Mechanical Engineering, Cranfield Institute of Technology, Cranfield, Bedford

SYNOPSIS

Numerical studies of three-dimensional viscous flows are now capable of predicting the complex distorted flows occuring in ducting systems which can lead to reductions in diffuser efficiency, fan stall margins and fan efficiency. To achieve this predictive capability the fully elliptic Navier-Stokes equations, incorporating the k-ε turbulence model, are solved in a generalised coordinate system. An efficient solution procedure based on the SIMPLE algorithm for velocity-pressure coupling and a method for its implementation on a non-staggered grid arrangement are described. The computer code used is capable of handling irregularly shaped ducted flow domains and non-uniform inlet conditions. Results are presented for the flow in straight and curved ducts and diffusers of a rectangular cross-section. Results are compared with experimental data and show the capability of the present method to reveal the flow patterns and to provide good agreement between predicted and measured values.

## NOMENCLATURE

| | | |
|---|---|---|
| $A, A^P$ | : | Coefficients of the finite difference equations. |
| $C_\mu, C_1, C_2$ | : | Constants in k-ε model |
| $D$ | : | Width of the square duct cross section. |
| $G$ | : | Rate of production of turbulence kinetic energy. |
| $g^{ij}$ | : | Grid metric coefficients. |
| $J$ | : | Jacobian of the coordinate transformation. |
| $k$ | : | Turbulence kinetic energy |
| $p$ | : | Static pressure. |
| $p'$ | : | Correction to the static pressure. |
| $S$ | : | Source term in general transport equation. |
| $u, v, w$ | : | Cartesian velocity components. |
| $U, V, W$ | : | Curvilinear velocity components. |
| $x, y, z$ | : | Cartesian coordinates. |
| $a$ | : | Half width of the square duct cross section. |
| $\Gamma$ | : | Diffusion coefficient for general transport equation. |
| $\varepsilon$ | : | Dissipation rate of turbulence energy. |
| $\xi, \eta, \zeta$ | : | Curvilinear coordinates. |
| $\mu$ | : | Laminar viscosity. |
| $\rho$ | : | Density. |
| $\sigma_k, \sigma_\varepsilon$ | : | Effective Prandtl numbers for k and ε. |
| $\Phi$ | : | General scalar quantity. |
| $\Delta\xi, \Delta\eta, \Delta\zeta$ | : | Cell boundary dimensions in the transformed plane. |

## Subscripts.

| | | |
|---|---|---|
| $P$ | : | Control grid point. |
| $E, W, N, S, F, B$ | : | Neighbouring grid points. |
| $i, j, k$ | : | Coordinate direction indices. |
| $T$ | : | Turbulent. |
| $a$ | : | General neighbouring grid point. |
| $\Phi$ | : | Related to scalar $\Phi$. |

| | | |
|---|---|---|
| $x, y, z$ | : | Derivatives in respect to $x, y, z$. |
| IN | : | Inlet. |

## Superscripts.

| | | |
|---|---|---|
| $i, j, k$ | : | Coordinate direction indices. |
| $-$ | : | Value obtained by simple averaging. |

## 1. INTRODUCTION

The matching of the fan and its ducting system in the installation design phase is important as the design of the inlet connections can have a significant effect on fan performance. This is because non-uniform flow at the inlet can lead to fan performance penalties and alter the pressure developed. Diffusers also represent key elements in the system and the success of their design is essential for the stable, low-loss and low noise systems.

The design of the ducting systems is usually based on empirical correlations and rules which attribute to each part of the system a pressure loss expressed as a fraction of the local dynamic head, Miller [1]. Usually the mechanisms involved, however, are far more complex especially in bends where the existing pressure gradient due to curvature causes secondary flows which introduce additional loss, modify the axial velocity profile and can produce swirl.

For a number of decades much of the effort devoted to the development of installation systems has been experimental, using a 'build and test' approach. As the power of modern computers has increased and cheaper computer facilities have become available, the 'numerical' or 'computational' approach has become increasingly attractive with considerable potential for the future as, although testing remains important, the trend is toward greater dependence on computational predictions. If adequately configured, computer studies can be used to produce the optimum design with reduced testing programs.

In recent years the 'pressure correction' technique has been used for the computation of flows in ducts. The method is used to solve the mass-averaged Navier-Stokes equations, which are expressed in a primitive variable formulation, by calculating the pressure through successive corrections in order that the solution of the momentum equations can be arranged (by successive trials) to satisfy the continuity equation.

For cases in which the main flow direction does not change, the flow equations can be parabolised by neglecting the diffusive terms in the through-flow direction. In this 'purely parabolic' formulation, presented by Patankar and Spalding [2], the flow conditions are transferred downstream only and consequently as in two-dimensional boundary layer calculations, the solution can be obtained with one iteration of the forward marching procedure. Several researchers have reported procedures for this class of flows, for example Ghia and Sokhey [3].

In many flow situations, however, the downstream conditions influence the upstream flow (for instance in ducts with large curvature where the pressure effect is transmitted upstream). A solution approach to this class of flows which are called 'partially parabolic' was presented by Pratap and Spalding [4], who developed a multipass algorithm for establishing a three-dimensional pressure field. This procedure was expanded by Moore and Moore [5], who arranged the equations in orthogonal curvilinear coordinates and thereby the method could be used to analyse the flow through complex geometrical shapes. This method used a two-dimensional pressure correction on each cross-stream plane which was not coupled to the streamwise pressure gradient unlike the procedure of Pratap and Spalding. A three-dimensional correction was employed, however, after each complete pass of the marching procedure. Rhie and Chow [6], used the pressure correction technique to solve two-dimensional flow fields in general curvilinear coordinates employing a non-staggered grid arrangement on which the variables were stored and calculated on the same grid location, unlike the conventional staggered grid approach used in the other models researched, Caretto et al [7]. Rhie [8] expanded this technique to three-dimensional flows having as a basis the partially parabolic procedure of Pratap and Spalding [4].

This paper describes a computational method developed, by the present authors, at Cranfield Institute of Technology, based on the work of Lapworth [9], Lapworth and Elder [10], for the prediction of three-dimensional, incompressible, viscous, turbulent flows in ducting systems of arbitrary shape. The method is a finite volume pressure correction SIMPLE-based algorithm, in which the pressure field is determined by successive corrections in order that the velocities satisfy the global and local continuity condition. The approach is fully elliptic and can handle regions with separated and recirculating flows. The equations use the Cartesian velocity components and the pressure as dependent variables and a strong conservation

form of the equations is used as they are expressed in a generalised coordinate system. For the simulation of the turbulence effects the k-ε model of turbulence, Launder and Spalding [11], is used. The computation of such three-dimensional flows requires a large computer memory and considerable 'running' time with a large number of grid points which must be used to facilitate the proper resolution of the viscous effects (losses and boundary layer phenomena).

This paper describes results obtained by applying this computer code to the prediction of the flow in three configurations. The first involves the developing turbulent flow in a straight duct of square cross-section, the second, the turbulent flow in a curved duct of rectangular cross-section, and finally, the turbulent flow in a low-aspect ratio diffuser of rectangular cross-section. The predicted values are compared with available results in the literature and, in general, good agreement is achieved.

## 2. MATHEMATICAL MODELLING

### 2.1 Governing Equations

For the purpose of giving the necessary background, the basic formulation is briefly discussed below. Because of the low airflow speeds in ducting systems, the flow is assumed to be incompressible. The steady-state turbulent flow is governed by the time averaged conservation equations of mass and momentum (the energy equation is only required if the flow is compressible). In a Cartesian coordinate system these equations can be expressed as follows:

Continuity

$$\frac{\partial(\rho u_j)}{\partial(x_j)} = 0 \qquad (1)$$

Momentum equations

$$\frac{\partial(\rho u_j u_i)}{\partial x_j} = \frac{\partial}{\partial x_j}\left[(\mu + \mu_T)\frac{\partial u_i}{\partial x_j}\right]$$

$$+ \frac{\partial}{\partial x_j}\left[(\mu + \mu_T)\frac{\partial u_j}{\partial x_i}\right] - \frac{\partial p}{\partial x_i} \quad (i = 1,2,3) \quad (2)$$

where $\mu$ is the laminar viscosity; $\mu_T$ is the turbulent viscosity; and there is a summation over all repeated indices.

The turbulent viscosity $\mu_T$ which is introduced in order that the effects of turbulence on the flow can be taken into account, is calculated by using the standard k-ε model, Launder and Spalding [11], in which the turbulent viscosity $(\mu_T)$ is related to the turbulence kinetic energy, k, and the rate of dissipation of turbulence energy, ε, by:

$$\mu_T = C_\mu \rho k^2/\varepsilon \qquad (3)$$

where: $C_\mu$ is a constant given in Table 1 and k and ε satisfy the scalar transport equations:

$$\frac{\partial(\rho u_j k)}{\partial x_j} = \frac{\partial}{\partial x_j}\left[\frac{\mu_T}{\sigma_k}\frac{\partial k}{\partial x_j}\right] + G - \rho\varepsilon \qquad (4a)$$

$$\frac{\partial(\rho u_j \varepsilon)}{\partial x_j} = \frac{\partial}{\partial x_j}\left[\frac{\mu_T}{\sigma_\varepsilon}\frac{\partial k}{\partial x_j}\right] + (C_1 G - C_2 \rho\varepsilon)\frac{\varepsilon}{k} \qquad (4b)$$

G is the rate of production of turbulent kinetic energy. The empirical constants arising in the k–ε model are given the values found in Table 1.

Table 1

| $C_\mu$ | $C_1$ | $C_2$ | $\sigma_k$ | $\sigma_\varepsilon$ |
|---------|-------|-------|------------|---------------------|
| 0.09    | 1.47  | 1.92  | 1.0        | 1.3                 |

In order to minimize the computer storage and run times, the dependent variables at the wall are linked to those at the first grid node from the wall by equations which are consistent with the logarithmic–law of the wall and are known as wall functions [11].

All the preceeding Cartesian conservation equations for mass, momentum and turbulence scalars are elliptic partial differential equations and can be expressed in the form of a general transport equation for an arbitrary dependent variable $\Phi$:

$$\frac{\partial(\rho u_j \Phi)}{\partial x_j} = \frac{\partial}{\partial x_j}\left[\Gamma\frac{\partial\Phi}{\partial x_j}\right] + S(x,y,z) \qquad (5)$$

where $\Gamma$ is the effective diffusion coefficient and S is the source term. By expressing the equations in the form of the general equation (5), a unique calculation procedure is adequate for the solution of these equations.

## 2.2 Transport equations in a generalised coordinate system

In general, the geometry of the ducting system does not conform to the rectilinear coordinate system. The use of body–fitted coordinate system is desirable to allow proper treatment of the near–wall region as well as to reduce the skewness between the streamlines and the coordinate lines. The numerical grid used in the current work is a curvilinear mesh where the curvilinear coordinate lines coincide with the boundaries at the physical domain. With this approach, a coordinate transformation is necessary which maps an irregularly shaped three–dimensional physical domain into a three dimensional computational domain having a regular shape, where well–established finite volume techniques can be applied to solve the problem. If new independent geometric variables $\xi$, $\eta$ and $\zeta$ are introduced, the form of the general transport equation (5), is changed according to the general transformation $\xi=\xi(x,y,z)$, $\eta=\eta(x,y,z)$ and $\zeta=\zeta(x,y,z)$, to the following form:

$$\frac{\partial(\rho U_j \Phi)}{\partial \xi_j} = \frac{\partial}{\partial \xi_j}\left[J\Gamma g^{jk}\frac{\partial\Phi}{\partial \xi_k}\right] + J\cdot S(\xi,\eta,\zeta) \qquad (6)$$

where J is the Jacobian of the transformation, $U_j =(U,V,W)$ are contravariant velocity components scaled by J:

$$U = J\cdot(\xi_x u+\xi_y v+\xi_z w) \qquad (7a)$$

$$V = J\cdot(\eta_x u+\eta_y v+\eta_z w) \qquad (7b)$$

$$W = J\cdot(\zeta_x u+\zeta_y v+\zeta_z w) \qquad (7c)$$

$g^{ij}$ are the components of the metric tensor:

$$g = \frac{\partial\xi_i}{\partial x_k}\cdot\frac{\partial\xi_j}{\partial x_k} \qquad (7d)$$

and $S(\xi,\eta,\zeta)$ is the transformed source term. In order to derive the above equation (6) the partial derivatives found in equation (5) are transformed using the chain rule in a fully conservative form; for example the transformation of the partial derivative of $\Phi$ with respect to x is given by:

$$\frac{\partial\Phi}{\partial x} = \frac{1}{J}\left[\frac{\partial(J\xi_x \Phi)}{\partial\xi} + \frac{\partial(J\eta_x \Phi)}{\partial\eta} + \frac{\partial(J\zeta_x \Phi)}{\partial\zeta}\right] \qquad (7e)$$

## 3. METHOD OF COMPUTATION

### 3.1 General transport equation

For the purpose of solution, the flow domain is overlaid with a cuvilinear grid whose intersection points are the locations at which all variables are stored and calculated. Around each grid point small control volumes are constructed, figure 1. In the present study the faces of each control volume are posed at locations midway between the corresponding neighbouring nodes.

The governing flow equations in the generalised system are intergrated over the control volumes and evaluated eventually by using finite difference approximations. The necessary quantities on the cell boundaries are estimated by using linear interpolation between adjacent grid points or obtained from upstream values.

The finite difference equations may be written as relations between the values of $\Phi$ at P and its values at the neighbouring grid points:

$$A_P \Phi_P = A_E \Phi_E + A_W \Phi_W + A_N \Phi_N + A_S \Phi_S$$
$$+ A_B \Phi_B + A_F \Phi_F + J\cdot(S_\Phi)\Delta\xi\Delta\eta\Delta\zeta \qquad (8)$$

where the coefficients A's involve density, convection, diffusion and geometrical quantities of the control cells. These coefficients are calculated by applying the 'hybrid' scheme, Patankar [12], which combines the convective and diffusive terms according to the magnitude of the grid cell Reynolds number by using either a central or an upwind differencing scheme. Upwind differencing introduces numerical diffusion but avoids numerical instability associated with the centered differencing of the convection term.

### 3.2 Solution of the pressure field

The SIMPLE (Semi–Implicit Method for Pressure–Linked Equations) Algorithm , [2],

[7], [12], is utilised for the solution of the pressure field, whereby the pressure field is gradually established through corrections which promote the satisfaction of the global and local continuity.

In the present study, a non-staggered grid arrangement is used in which all the variables are stored and calculated at the same grid nodes, as opposed to a staggered grid arrangement employed by many researchers. The pressure oscillations arising in a non-staggered grid are eliminated using the pressure correction scheme proposed by Rhie and Chow [6].

The pressure field is established through a three-step correction procedure. Details of this procedure can be found in Lapworth and Elder [10].

In the first step, a one-dimensional global correction to the streamwise velocity and pressure is implemented at each cross-stream plane. The uniform correction to the streamwise velocity component is obtained by applying global continuity and the corresponding one-dimensional correction to the pressure is obtained from the approximate streamwise momentum equation. This procedure accelerates the establishment of the correct pressure field.

In the second step, a two-dimensional elliptic correction is performed to the pressure and cross-stream velocities to satisfy the local continuity at the control volumes surrounding each grid node in the cross-stream plane. By using the discretised continuity equation, an equation similar to equation (8), for the pressure correction p' is derived and is solved two-dimensionally. The values of pressure correction p' obtained, are used to correct the cross-stream velocities U and V and to update the pressure.

In the third step a three dimensional correction to the pressure field alone is executed. This procedure accelerates the transmission of the elliptic pressure effects.

4.    SOLUTION PROCEDURE

A space marching procedure is implemented from the inlet to the outlet of the flow (in the primary flow direction).For the initiation of this procedure, the velocity components and the turbulent parameters k and ε are required at the inlet plane. The entire solution algorithm can be summarized as follows:

1.    The solution of the momentum equations with a given pressure field yields values for u,v and w.

2.    The global pressure correction is applied. The local continuity is enforced and the pressure and the corresponding velocities are updated accordingly. A three-dimensional pressure correction accelerates the establishment of the correct pressure field.

3.    The equations (4a) and (4b) for the k and ε are solved and the turbulent viscosity is calculated by using equation (3).

4.    The computational domain is shifted to the next cross-stream plane and the process from 1 to 4 is repeated.

The one pass completed through the steps 1 to 4 is repeated until a convergent solution is achieved.

The solution of the algebraic system of equations for the velocity components, the pressure, the pressure correction and the turbulence scalars, at each cross plane, is carried out by applying the Alternating-Direction Implicit (ADI) algorithm with the use of the Tri-Diagonal-Matrix-Algorithm (TDMA), Patankar [12].

5.    COMPUTATIONAL RESULTS

5.1    Developing turbulent flow in a straight duct of square cross-section

As a first test case the developing turbulent flow in a straight duct of square cross-section is simulated. The turbulent flow in a straight duct of non-circular cross-section is characterized by the presence of transverse mean secondary flow even in the fully developed flow regions. This secondary flow defined by Prandtl as secondary flow of the second kind, is a result of Reynolds stress gradients in the corner region. Although the magnitude of the secondary velocity is only a small percentage (2-3%) of the streamwise bulk velocity, it exerts a significant influence on the global and local properties of the flow. The contours of the axial mean velocity bulge outwards near the corners as a result of the transport of high-momentum fluid from the core fluid to the corner regions through the circulatory transverse flow.

Although experimental investigations on fully developed flows in non-circular ducts were conducted by many workers, available experimental data are scarce for the case of developing turbulent flow. Gessner, Po and Emery [13], reported a thorough experimental investigation on the three dimensional developing flow in a square duct at a Reynolds number of 250000. Melling and Whitelaw [14], performed Laser Doppler measurements at two axial stations of the square duct at a Re=42000. The experimental data from both the above investigations have been used by many workers especially for the validation of turbulence models and are used for comparisons with the present predictions.

It is generally accepted that a turbulence model using the assumption of an isotropic eddy viscosity , as the k-ε model, can not predict the secondary motion and a more refined modelling of the turbulent stresses is necessary, Launder and Ying [15]. At this stage, however, only results obtained with the use of k-ε model are presented in order to be consistent with the mathematical formulation presented above although the Launder-Ying model has been already incorporated in the computer code.

Figure 2a shows the predicted axial centreline velocity compared with the experimental data obtained by Gessner et al. There is an increase in the centreline velocity as far as z/D=40 as a result of the boundary layer growth on the walls confining and accelerating the core fluid. Subsequently the centreline velocity decreases as a result of the redistribution of the momeentum across the duct through the secondary circulatory motion. The predictions agree with the measurements quite well in the initial part of the duct but indicate their peak value at $z/D=35$. After that it tends asymptotically to the value measured for the fully turbulent flow. The corresponding predictions for the experiment carried out by Melling and Whitelaw are shown in figure 2b. The centreline velocity presents a trend similar to that of the previous experiment. The peak of the ratio W/Wbulk occurs at $z/D=25$ and it becomes almost constant before it falls gradually.

The boundary layer development along the duct for the flow conditions of Gessner, Po and Emery [13], can be observed in figure 3. The axial velocity profiles are plotted against the distance from the wall bisector. The gap between the predicted values and the experimental data is greater at the $z/D=24$ and $z/D=40$ stations. This occurs due to the insufficiency of the k-ε model to simulate the turbulence-driven secondary flows and subsequent bulging of the axial velocity contours outwards to the corner bisector and inwards to the wall bisector.

The grid employed for the simulation of the experiment performed by Gessner et al. had 14 by 14 nodes covering one quarter of the cross-plane because of the symmetry of the flow and 48 planes covering 60 diameters length of the duct. The computations were performed on a VAX 8650 computer system and reached convergence after 12955 sec of CPU time or 0.0025 sec per grid point per iteration.

## 5.2 Flow in a curved passage of rectangular cross-section

In order to check the capability of the presented computational method to treat strong curvature in the longitudinal direction, the turbulent flow in a 90° bend of rectangular cross-section was calculated.

This calculation simulated the experimental conditions reported by Pratap [16]. Figure 4 shows the configuration of the curved duct. The test section of the duct was rectangular and the height and the width were equal to 1.22m and 0.304m respectively. The duct curved through 90° with a centre line radius of 2.52m.

The flow domain for which computations have been carried out consisted of a 1.22m long straight section and the 90° of the curved duct. Only half of the width was considered because of the symmetry of the flow. The grid which was employed consisted of 14 nodes in the x, 16 nodes in the y and 34 nodes in the streamwise direction. The polar angle between the cross-sectional planes in the curved part was equal to 3.75°. The grid was not uniform with more points in the near-wall region.

The Reynolds number based on the hydraulic diameter of the duct and the mean velocity was 700000. The flow conditions at the inlet plane of the calculation domain were fixed according to the experimental measurments at the location 1.22 upstream of the curved part.

The predictions showed a complex three-dimensional flow pattern due to the secondary motion driven by the strong pressure gradient induced by the curvature. Two counter-rotating vortices were found to exist with a maximum velocity of the secondary motion about 20% of the maximum primary velocity. This secondary motion is shown, for the 75° plane, in figure 9 where only half of the duct is plotted. As a result of this secondary motion, the primary velocity profiles were distorted and showed a maximum towards the inside of the curved duct. Figures 5 and 6 show the streamwise velocity profiles at the 11.25° and the 33.75° planes for distances X=0.01m and X=0.02m from the side wall of the duct. The velocities were non-dimensionalised with the magnitude of the total velocity Q at the centre of the corresponding cross-sectional plane. The velocity shows a peak near the inner wall of the duct which occured due to the pressure gradient developed from the inner to the outer wall and was caused by the curvature of the bend. This pressure distribution is presented in figures 7 and 8 in terms of a Cp coefficient defined as:

$$Cp = \frac{p - p_c}{0.5 \cdot \rho \cdot Q^2}$$

where $p_c$ is the pressure at the centre of the cross-stream plane.

The Cp coefficient presents its maximum value at the outer wall and this value reduces towards the inner wall due to the action of the centrifugal forces. The results obtained are in a very good agreement with the measured values in most of the cross-sections of the duct.

The calculations were performed on a VAX 8650 computer system and convergent solution was obtained after 560 iterations of the marching procedure and took 14085 sec of CPU time corresponding to 0.0033 sec per grid point per iteration.

## 5.3 Flow in a low-aspect-ratio diffuser

To check the capability of the present analysis to treat flows with streamwise diffusion, the turbulent flow in a low-aspect-ratio, straight, rectangular diffuser was studied. McMillan and Johnston [17], [18], conducted an experimental investigation of air flows in low-aspect-ratio rectangular diffusers of different outlet to inlet area ratios and for different Reynolds numbers. Figure 10 shows the configuration of the diffuser.

The present calculation simulated the experimental conditions reported by McMillan and Johnston, [18], for an unstalled diffuser of an area ratio equal to 2.1:1 and with a fully developed turbulent flow at the inlet of the diffuser.

The computational inlet plane was located $0.8*W_1$ lengths upstream of the diffuser inlet section following the calculations performed by Rhie [8], for the same test case. Only one quarter of the whole diffuser configuration was considered in the computations because the flow was symmetrical in both the cross-stream directions. A grid of 25 by 10 by 52 was used. Figure 11 shows the static pressure distribution along the diffuser for Re=20600 and Re=50600. The local pressure recovery is expressed in terms of a Cp coefficient defined as :

$$Cp = \frac{p - p_{IN}}{0.5 \cdot \rho \cdot W_{IN}^2}$$

where p is the local static pressure, $p_{IN}$ is the inlet static pressure and $W_{IN}$ is the area-averaged inlet velocity. The Cp coefficient increases along the diffuser with higher values for the higher Reynolds number. Generally the predicted values are in good agreement with the corresponding measured data. The necessary CPU time was 13282 sec on a VAX 8650 computer system.

6.    CONCLUSIONS

A calculation procedure has been developed to compute the incompressible viscous flow through arbitrary three-dimensional duct configurations. The full time-averaged Navier-Stokes equations have been solved using an iterative pressure correction method. The k-ε turbulence model has been incorporated to simulate the effects of turbulence on the mean flow.

The application of the method for the prediction of the developing flow in a square duct has been demonstrated, although the inadequacy of the k-ε model to capture the turbulence driven secondary flows in the corner regions of the duct has also been observed. The pressure driven secondary flows and their influence on the streamwise velocity profiles have been modelled accurately for the case of a 90° curved duct of rectangular cross-section. The pressure recovery in a low-aspect-ratio diffuser has been well predicted and overall agreement with the measured values has been demonstrated.

The method can be used to represent a range of flow conditions with acceptable accuracy for many ducting configurations. Further developments of the presented computational method include the simulation of the compressibility effects for subsonic but high Mach number flows. Also, further applications can include parametric studies of the effects of inlet conditions and various geometrical parameters on the performance of axial and centrifugal fans by simulating the flow through the rotating passages of the fan.

REFERENCES

1.    Miller D.S.
      'Internal Flow Systems'
      BHRA, Fluid Engineering, 1978.

2.    Patankar S.V. and Spalding D.B.
      'A Calculation Procedure for Heat, Mass and Momentum Transfer in Three-Dimensional Parabolic Flows'
      Int. J. Heat Mass Transfer, Vol 15, pp 1787-1806, 1972.

3.    Ghia K.N. and Sokhey J.S.
      'Laminar Incompressible Viscous Flow In Curved Ducts of Rectangular Cross Sections'
      J. of Fluids Engg., Vol 99, pp 640-648, 1977.

4.    Pratap V.S. and Spalding D.B.
      'Fluid Flow and Heat Transfer in Three-Dimensional Duct Flows'
      Int. J. Heat Mass Transfer, Vol 19, pp 1183-1188, 1976.

5.    Moore J. and Moore J.G.
      'A Calculation Procedure for Three-Dimensional, Viscous, Compressible, Duct Flow. Part 1 - Inviscid Flow Considerations'
      J. of Fluids Engg., Vol 101, pp 415-422, 1979.

6.    Rhie C.M. and Chow W.L.
      'Numerical Study of the Turbulent Flow Past an Aerofoil with Trailing Edge Seperation'
      AIAA Journal, Vol 21, No 11, pp 1525-1532, 1983.

7.    Caretto L.S., Gosman A.D., Patankar S.V. and Spalding D.B.
      'Two calculation procedures for Steady Three-Dimensional Flows with Recirculation'
      Proceedings of the Third International Conference on Numerical Methods in Fluid Dynamics, Paris, 1972.

8.    Rhie C.M.
      'A Three-Dimensional Flow Analysis Method Aimed at Centrifugal Impellers'
      Computers and Fluids, Vol 13, No 4, pp 443-460, 1985.

9.    Lapworth B.L.
      'Examination of Pressure Oscillations Arising in the Computation of Cascade Flow using a Boundary Fitted Coordinate System'
      Int. J. Num. Meth. in Fluids, Vol 8, pp 387-404, 1988.

10.   Lapworth B.L. and Elder R.L.
      Computation of the Jet-Wake Flow Structure in a Low-Speed Centrifugal Impeller'
      ASME paper No 88-GT-217, 1988.

11.   Launder B.E. and Spalding D.B.
      'The Numerical Computation of Turbulent Flows'
      Comp. Meth. in Appl. Mech. Engg., Vol 3, pp 269-289, 1974.

12.   Patankar S.V.
      'Numerical Heat Transfer and Fluid Flow'
      McGraw Hill, New York, 1980.

13.   Gessner F.B., Po J.K. and Emery A.F.
      'Measurements of Developing Turbulent Flow in a Rectangular Duct', Turbulent Shear Flows I (Edited by Durst, et al.)
      Springer-Verlag, 1979.

14.   Melling A. and Whitelaw J.H.
      'Turbulent Flow in a Rectangular Duct'
      J. of Fluids Engg., Vol 78, part 2, pp 289-315, 1976.

15.   Launder B.E. and Ying W.M.
      'Prediction of Flow and Heat Transfer in Ducts of Square Cross-section'
      Heat and Fluid Flow, Vol 3, No 2, pp 115-121, 1973.

16.   Pratap V.S.
      'Flow and Heat Transfer in Curved Ducts'
      Imperial College of Science and Technology Report No. HTS/75/25, 1975.

17.  McMillan O.J. and Johnston J.P.
     'Performance of Low – Aspect – Ratio
     Diffusers With Fully Developed Inlet
     Flows, Part I – Some Experimental Results'
     J. of Fluids Engg., Vol 95, pp 385–392,
     1973.

18.  McMillan O.J. and Johnston J.P.
     'Performance of Low – Aspect – Ratio
     Diffusers With Fully Developed Turbulent
     Inlet Flows'
     Stanford Univ., Thermosciences Div. Report
     PD–14, 1970.

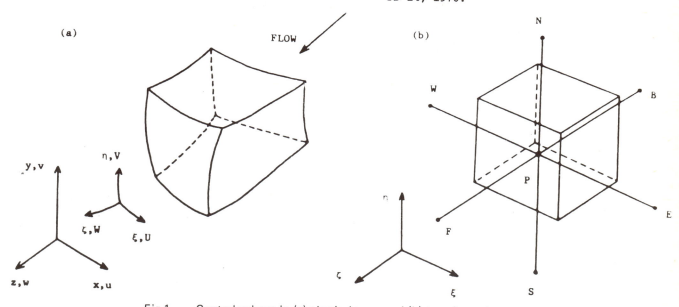

Fig 1    Control volume in (a) physical space and (b) transformed space

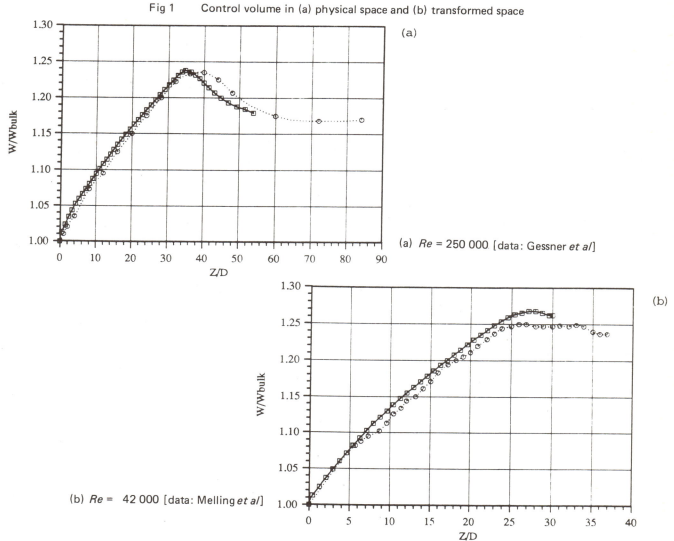

(a) *Re* = 250 000 [data: Gessner *et al*]

(b) *Re* =  42 000 [data: Melling *et al*]

Fig 2    Axial velocity distributions along the centreline
         of the duct

☐—☐    PRESENT PREDICTIONS USING THE k-ε MODEL

⊙····⊙    EXPERIMENTAL DATA

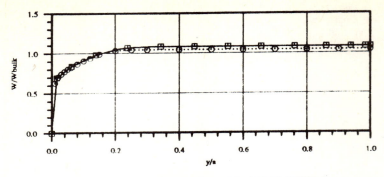

STREAMWISE VELOCITY DISTRIBUTION AT Z/D=8

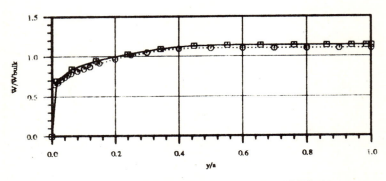

STREAMWISE VELOCITY DISTRIBUTION AT Z/D=16

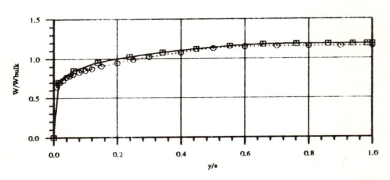

STREAMWISE VELOCITY DISTRIBUTION AT Z/D=24

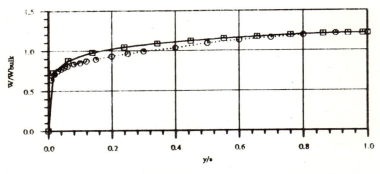

STREAMWISE VELOCITY DISTRIBUTION AT Z/D=40

⊞——⊞    PRESENT RESULTS USING THE k-e MODEL

⊙····⊙    EXPERIMENTAL DATA [ GESSNER,PO,EMERY 1979]

Fig 3    Axial velocity distributions along the wall bisector of the duct

© IMechE 1990 C401/013

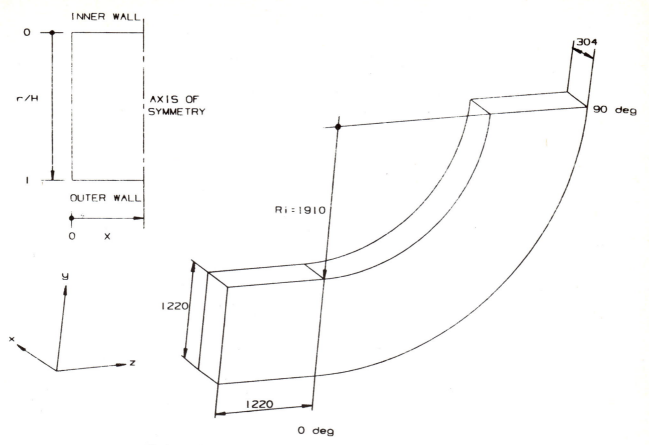

CROSS-STREAM PLANE

INNER WALL

0

r/H

AXIS OF
SYMMETRY

1

OUTER WALL

0    X

y

x

z

304

90 deg

Ri=1910

1220

1220

0 deg

Fig 4      Geometry of a 90° bend of rectangular cross-section

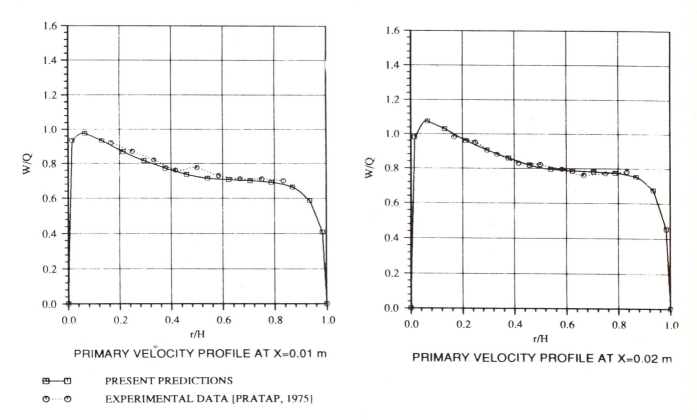

PRIMARY VELOCITY PROFILE AT X=0.01 m

PRIMARY VELOCITY PROFILE AT X=0.02 m

⊟——⊟        PRESENT PREDICTIONS

⊙····⊙        EXPERIMENTAL DATA [PRATAP, 1975]

Fig 5      Primary velocity profiles at the 11.25° plane of the curved duct

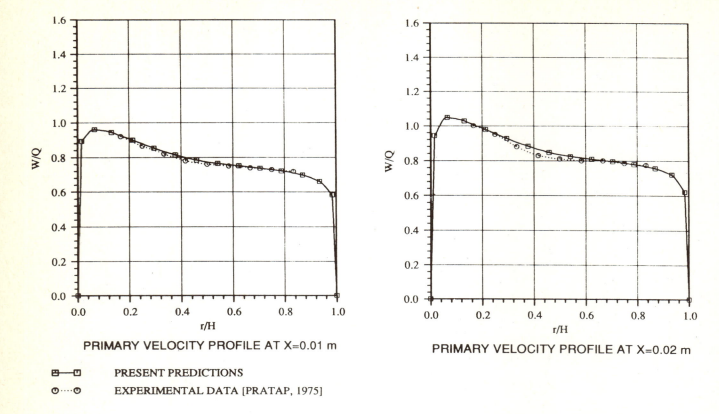

PRIMARY VELOCITY PROFILE AT X=0.01 m          PRIMARY VELOCITY PROFILE AT X=0.02 m

⊞—⊟          PRESENT PREDICTIONS
⊙····⊙          EXPERIMENTAL DATA [PRATAP, 1975]

Fig 6          Primary velocity profiles at the 33.75° plane of the curved duct

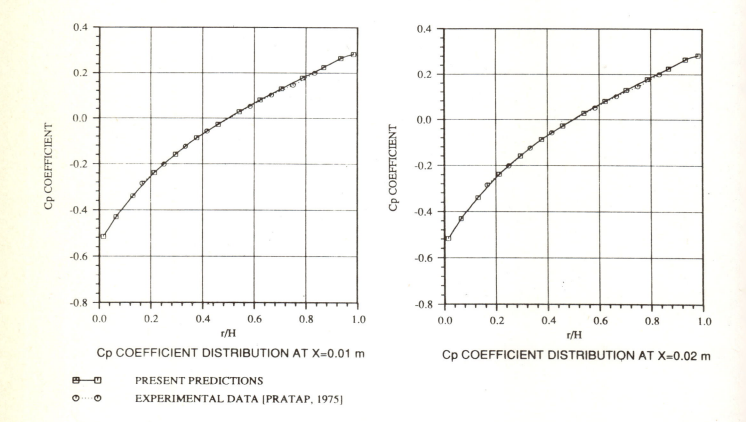

Cp COEFFICIENT DISTRIBUTION AT X=0.01 m          Cp COEFFICIENT DISTRIBUTION AT X=0.02 m

⊞—⊟          PRESENT PREDICTIONS
⊙····⊙          EXPERIMENTAL DATA [PRATAP, 1975]

Fig 7          Pressure distributions at the 11.25° plane of the curved duct

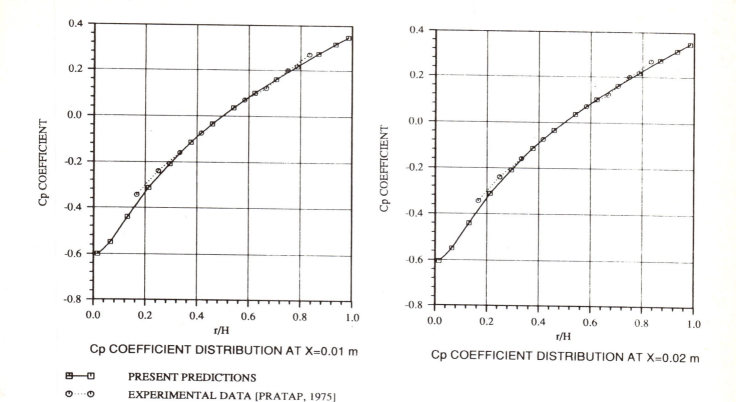

Cp COEFFICIENT DISTRIBUTION AT X=0.01 m

Cp COEFFICIENT DISTRIBUTION AT X=0.02 m

▣—▣    PRESENT PREDICTIONS

⊙····⊙    EXPERIMENTAL DATA [PRATAP, 1975]

Fig 8    Pressure distributions at the 33.75° plane of the curved duct

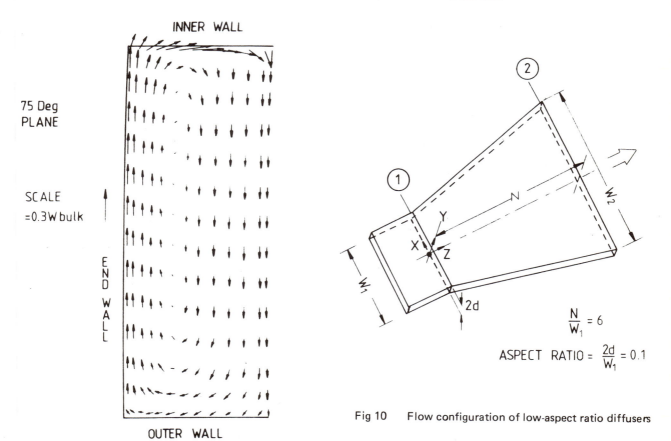

Fig 9    Secondary velocity vectors at the 75° plane of the curved duct

Fig 10    Flow configuration of low-aspect ratio diffusers

$$\frac{N}{W_1} = 6$$

$$\text{ASPECT RATIO} = \frac{2d}{W_1} = 0.1$$

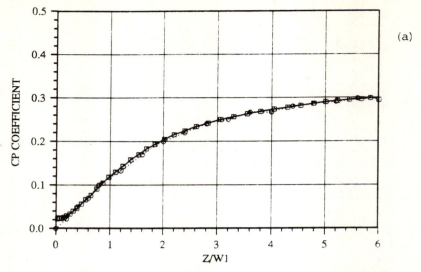

PRESSURE DISTRIBUTION ALONG THE DIFFUSER AT RE=20600

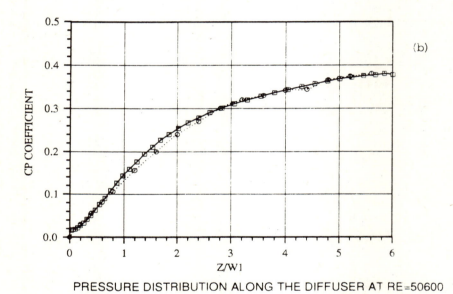

PRESSURE DISTRIBUTION ALONG THE DIFFUSER AT RE=50600

| | |
|---|---|
| ▣——▣ | PRESENT PREDICTIONS |
| ◉ · ·◉ | EXPERIMENTAL DATA [ McMILLAN, JOHNSTON 1970 ] |

Fig 11    Static pressure recovery for a low-aspect ratio diffuser at
(a) *Re* = 20 600 and (b) *Re* = 50 600

# Stall in low hub – tip ratio fans

**A B McKENZIE**, BSc, DTech, CEng, FIMechE and **H YU**, BSc
School of Mechanical Engineering, Cranfield Institute of Technology, Cranfield, Bedford

SYNOPSIS

Detailed flow fields were obtained in three axial fan builds with a hub–tip ratio of 0.32 by using both a conventional pneumatic probe and high frequency response pressure transducer probes. The time averaged results show that there was a low velocity region near the hub behind the rotor. The time resolved results indicate that rotating stall appeared within a very limited flow range, while axisymmetric stall dominated most of the stalled flow. The new features in low hub–tip ratio fans are presented and discussed. The effects of space–chord ratio, stators and on stalled flow were also studied.

NOMENCLATURE

| | |
|---|---|
| $Q$ | Yaw angle of the flow |
| $\rho$ | Density of the air |
| $\Phi$ | Flow coefficient |
| $\psi_{T-s}$ | Pressure coefficient |
| $\eta_T$ | Fan total efficiency |
| $p_s$ | Static pressure |
| $P_T$ | Total pressure |
| $T_0$ | The time period of rotating stall |
| $T$ | Time |
| $U_t$ | Rotor tip velocity |
| $V$ | Absolute velocity |
| $Va$ | Axial velocity |
| $V_w$ | Whirl velocity |

## 1. INTRODUCTION

When a fan or a compressor works at off design conditions the effect of unsteady flow can become significant and essential. Two phenomena often encou ntered at lower mass flow rate are stall and surge. Surge is associated with the overall compression system instability and it may cause large oscillations of mass flow and pressure rise. In industrial fans, however, surge rarely occurs because of their low rotational speed and small volume in the system. On the other hand, stall points or stall lines often limit the fan performance.

Stall may be divided into rotating stall and axisymmetric stall according to its kinematic behaviour. Rotating stall can be described as one or many distorted flow areas which propagate around the annulus at a fraction of rotor speed. Rotating stall leads to a large drop in pressure rise and efficiency and may make extra noise which cannot be accepted from the fan performance point of view. In addition, it may induce large vibrations of the blades and therefore it must be avoided for mechanical reasons.

Numerous research investigations have been carried out in order to understand rotating stall. With the aid of transient instrumentation, i.e. hot wires and fast response pressure transducers, and digital computers, it is possible not only to measure the basic features of rotating stall cells and their effects on the performance but also to obtain the detailed pressure and velocity fields inside the stall cells. One of the early works was that of Day (1). The experiments were carried out on an axial compressor with a hub–tip ratio of 0.8. The detailed measurements were obtained by using a slanted hot wire probe. The difference between part–span stall and full–span stall was assessed. It concluded that there are fairly large differences between the detailed flow field in single stage fans and multistage compressors.

Rotating stall cells in compressors can be full–span or part–span and the number of cells may also vary. Two full–span stall cells and up to eight part–span cells were observed in a single stage compressor with a hub–tip ratio of 0.78 by Breugelmans (2). Multi part–span cells located at the hub were reported by Giannissis (3) in a mismatched three stage compressor.

Although measurements in a low hub–tip ratio compressor were conducted in refs (4–7), the measurements were not in detail and there is not sufficient information to build up a clear picture of rotating stall in low hub–tip fans.

The experimental investigation presented here is to provide details of the unstalled and stalled flow in a very low hub–tip ratio industrial fan. The effects of pitch angle, space–chord ratio and the presence of an Outlet Guide Vane (OGV) section were studied with emphasis on the difference between the stalled flow in fans and compressors at off design conditions. Some important features of rotating stall in fans are described.

## 2. TEST FACILITY AND INSTRUMENTATION

The test rig and the blade profile are shown schematically in Fig. 1. The rig used consisted of a conical intake, a set of straighteners, inlet duct, the test fan, outlet duct and a downstream throttle which controlled the mass flow rate. All fans tested had the same outer diameter of 0.965m and the rotational speed of the fan was 1470 rpm. The motor was located downstream of the rotor except when the OGV was fitted upstream of the fan.

Three different fans were tested. Build 1 was the rotor only fan with 10 blades. The hub–tip ratio was 0.32 and the blade chord length at tip was 129mm. Three pitch angles 16, 24 and 30 degree were tested (the pitch angles were measured from the circumferential direction at the tip). The second build was the same rotor with an OGV section. In this case, the motor was located upstream of the rotor. The OGV section had 12 blades which were bent sheet metal with large camber and chord near the hub. Build 3 was the rotor with 5 blades. This build was undertaken in an attempt to assess the effect of space chord ratio on the stall performance.

For unstalled flow conditions, a conventional three hole pneumatic probe was employed to measure the flow angle, static pressure and total pressure. The probe was traversed at positions 88mm upstream of the rotor and 135mm downstream of the rotor. The last radial position is about 2.5 percent of the blade height away from the end wall.

A three–hole cylindrical probe with three fast response pressure transducers built into the probe, close of the ports, was used to detect the transient pressure signals when the fan operated in rotating stall conditions, Giannissis (3). Before the signals were fed into an on–line computer which performed data acquisition and processing, they passed through a bank of amplifiers, low pass filters and a sample and hold unit with two trigger circuits. Because of the unsteady nature of the signals, phase–lock sampling and ensemble averaging techniques were necessary. The reference signal was obtained by using a total pressure probe fixed upstream of the rotor. The sampling pulses were generated from a slotted disk mounted on the rotor. The output from the probe could be sampled at 36 points in each rotor revolution. For each position, a total of 100 samples were taken. The digitized data were recorded on floppy disk and then transferred to the main computer for further data analysis. The velocity and pressure of the flow could be calculated from the ensemble averaged results combined with the probe calibration data.

Preliminary results indicated that the flow angle varied about 120 degrees ahead of the rotor and to achieve the required angular range the three hole probe was offset by 70 degrees. Behind the rotor, the probe was set at 60 degrees as the variation of flow angles were found to be relatively small and 60 degrees was close to the mean flow angle.

## 3.    EXPERIMENTAL RESULTS

### 3.1    Overall fan performance

Fan static pressure rise characteristics were obtained according to British Standards 848. The fan static pressure is defined as fan outlet static pressure at plane 3 minus fan inlet total pressure at plane 2, Fig. 1. The mass flow measurements were obtained from static pressure taken in the calibrated conical mouth intake. The pressure rise was normalized to rotor tip dynamic head, i.e. $\frac{1}{2}\rho U_t^2$. The flow coefficient represented the average axial velocity in the duct divided by the rotor tip velocity. The fan efficiency was calculated using the fan total pressure.

The characteristics of Build 1, which is the isolated rotor with ten blades, are shown in Fig. 2. The fan static pressure at three different pitch angles are shown. For the 16 degree pitch angle, the characteristic is continuous and the slope of the curve is always negative. No rotating stall was found. However, when the mass flow was very small, reverse flow was found near the blade tip upstream of the rotor. The reverse flow can be classified as axisymmetric stall since the reverse flow is continuous around the whole annulus at the tip. If the throttle is closed further, the reverse flow region increases in radial extent, but the flow pattern does not become circumferentially distorted. For the 24 degree pitch angle, the characteristic is quite similar to that of the 16 degree pitch angle. Before the axisymmetric stall occurred, however, the pressure rise peaked and a region with positive slope was detected. Again, there was no occurrence of rotating stall.

With the 30 degree pitch angle, a part–span stall cell was detected after the stall point. The rotating stall existed only in a very limited range. As the slope of the characteristics becomes negative, the rotating stall changes into axisymmetric stall and the pressure rise increased with closing of the throttle. When opening the throttle, a hysteresis loop is found and the stall cessation mass flow is bigger than its inception.

It will be observed in Fig. 2 that the total efficiency curves are very flat for the higher pitch angle configurations compared with that of lower pitch angle. The peak efficiency are around 74%. When rotating stall occurred, the efficiency dropped about 12 percentage points.

The fan static pressure characteristics of Builds 2 and 3 are plotted in Fig. 3. For Build 2, which is Build 1 plus the OGV, it is clear that the OGV does not improve the pressure rise in this particular case. Build 1 and 2 stall at nearly the same flow whereas for Build 3 the fan static pressure curves are flatter and the stall flow is reduced significantly. This appears because the space chord ratio has to be a large influence on the blade loading and eventually on stall.

When a fan operates in stall, the wall static pressure measurement can be affected seriously, influencing the derived characteristics. In Fig. 4 the characteristics are re–calculated using atmosphere pressure (instead of that measured just upstream of the fan) as the inlet total pressure. In the steady flow region, the shape of the characteristics do not change significantly. For stalled flow, however, the slope of the characteristics are significantly modified tending to have a more negative slope. The shut–off pressure rise, i.e. the pressure rise at zero mass flow, becomes the peak pressure for all characteristics. For Build 1 and 2, the shut–off pressures are around 0.355. It is clear that the shut–off pressure is independent of blade pitch angle and also the presence of the OGV. It is important, however, to note that the shut–off pressure reduces significantly for Build 3, indicating that the shut–off pressure is a function of blade number (or space chord ratio).

## 3.2 Time averaged traversing results

Detailed flow fields were measured ahead of, and behind, the rotor by employing a three hole pressure probe. The results were obtained for all three builds and various pitch angles. Since the amount of data is large, it is impossible to show all results here. After analyzing the data carefully, it was found that, though there exists some minor difference, there are several common features which tend to dominate, hence Build 1 with the 24 pitch angle setting will be presented as an example.

The measurements for this build, downstream of the rotor, are illustrated in Fig. 5. The most important feature is the low energy region near the hub. From the axial velocity profiles, it can be seen that the axial velocity begins to decrease at about 40 percent of the span. Very low velocity, even reverse flow was found within 20 percent of span from the hub. The reverse flow region, however, has little effect on the stability of the fan as a whole and did not appear to induce instabilities such as rotating stall.

In order to illustrate the low flow region more clearly, a sketch of the stream lines for three different flow coefficients is plotted in Fig 6. It clearly indicates that the stream lines shift outward to the casing with reducing flow coefficient. This picture also suggests that the centrifugal effect is remarkably strong in low hub–tip ratio fans.

## 3.3 Unsteady measurements

Flow measurements in the rotating stall regime were taken for all three builds at 30 degree pitch angles. Because the rotating stall existed only within a very limited flow range, detailed measurements were only carried out at one flow condition for each build. A wall static pressure transducer probe was used as a reference to measure the number of stall cells and the speed of the cell. The general features of rotating stall cells are given in Table 1. For the fans tested, the stall cells are part–span single cell. It can be seen from Table 1 that the OGV had little influence on the speed of the cell. However, the size of the stalled flow area increases for the OGV build. On examining Build 3, the speed of the rotating stall is slower than Build 1 with more blades.

Fig. 7 shows the typical flow parameters, namely, the yaw angles, the total pressure coefficient, the static pressure coefficients, axial coefficients and the whirl velocity coefficients at the inlet and outlet of the rotor. It is seen that the flow angles ahead of the rotor are above 90 degree in the stalled area and the flow in the unstalled area is approximately axial. Behind the rotor, the flow direction has a much smaller variation. The axial velocity measurements also indicate the reverse flow in the stall cell upstream of the rotor. The radial traverse measurements were conducted at twelve radial positions upstream and downstream of the rotor. By combining all these data, the contours of yaw angles and total pressure coefficients ahead of the rotor are shown in Fig. 8(a) and 8(b) respectively. It is clear that the stall cell is located near

the casing and covers about 20 percent of the blade span. For the rest of the area, the flow is less disturbed. The yaw angle changes very sharply near the boundary of stalled and unstalled flow while the total pressure various smoothly from unstalled region to the centre of the stall cell. The total pressure has its maximum in the centre of the stall cell.

### TABLE 1
### CHARACTERS OF ROTATING STALL

|  | BUILD 1 | BUILD 2 | BUILD 3 |
|---|---|---|---|
|  | $\phi = 0.187$ | $\phi = 0.172$ | $\phi = 0.182$ |
| TYPE | PART-SPAN | PART-SPAN | PART-SPAN |
| SPEED Uc/Ut | 53 % | 53 % | 43 % |
| NO. OF CELLS | 1 | 1 | 1 |
| SIZE | 68 % | 82 % | 63 % |

**\* BLADE SETTING ANGLE = 30**

## 4. DISCUSSION OF RESULTS

Two modes of stalled flow were found in the low hub–tip ratio industrial fans tested. The rotating stall existed only over a very limited flow range while the axisymmetric stall dominates most of stalled flow. For the low pitch angle settings, the rotating stall did not exist. It is worth noting that axisymmetric stall gave continuous performance curves and therefore it would be preferred to rotating stall.

The static pressure rise across the rotor can be achieved by diffusing relative velocity or by use of centrifugal force. The large radial stream line shift in the fans tested indicated that centrifugal force plays a very important role in producing the static pressure rise near the hub. When designing a low hub tip ratio fan, the radial stream line displacement should not be ignored.

Considering the rotating stall inception, the conventional two dimensional rotating stall model cannot explain the results in fans. In the fans tested, the hub region always appears to be in difficulties with large losses and reverse flow. Rotating stall, however, is not triggered by this hub condition and only occurred when the blade tip or casing stalled. The impact of the strong centrifugal effect has not been included in any stall models to date. This could be the reason why all the models fail when applied to low hub–tip ratio fans.

## 5. CONCLUSIONS

Based on the results of this investigation, the following conclusions may be drawn.

(1) Two kinds of stall modes were found, rotating stall and axisymmetric stall. Rotating stall could only be detected over a very small flow range while axisymmetric stall dominates most of stalled flow range beyond the peak of the pressure rise characteristic.

(2) The shut-off pressure rise is dependent upon the number of rotor blades or space chord ratio.

(3) A large radial stream line shift was found in this low hub tip ratio fan and a retarded flow region or reverse flow region was measured at the hub.

(4) The rotating stall influences the upstream flow but it has less influence downstream.

## REFERENCES

1. DAY, I.J. Detailed Flow Measurements During Deep Stall in Axial Compressors. AGARD CP-177, 1977.

2. BREUGELMANS, F.A.E., MATHIOUDAKIS, K., and CASALINI, F. Flow in Rotating Stall Cells of a Low Speed Axial Flow Compressor. 6th ISABE, Paris, 1983.

3. GIANNISSIS, G.L., McKENZIE, A.B., and ELDER, R.L. Experimental Investigation of Rotating Stall in a Mismatched Three Stage Axial Compressor. ASME 88-GT-205, 1988.

4. SOUNDRANAYAGAM, S., and BALAKRISHNAN, K. A Model of Axial Impeller Stall. AIAA, pp 665-675, 1983.

5. TANATA, S., and MURATA, S. On the Partial Flow Rate Performance of Axial-flow Compressor and Rotating Stall – 1st Report, Influences of Hub-Tip Ratio and Stators. Bulletin of the JSME, Vol. 19, No. 117, March 1975.

6. TANATA, S., and MURATA, S. On the Partial Flow Rate Performance of Axial-flow Compressor and Rotating Stall – 2nd Report, Influences of Impeller Load and a Study of the Mechanism of Unstable Performance. Bulletin of JSME, Vol. 18, No. 117, March 1975.

7. DUNHAM, J. Observations of Stall Cells in a Single Stage Compressor. A.R.C. CP 589, 1961.

## ACKNOWLEDGEMENT

The authors would like to express their gratitude to Woods of Colchester for supplying the rig and to W.R. Woods-Ballard, T.W. Smith and I.R. Kinghorn for their valuable comments.

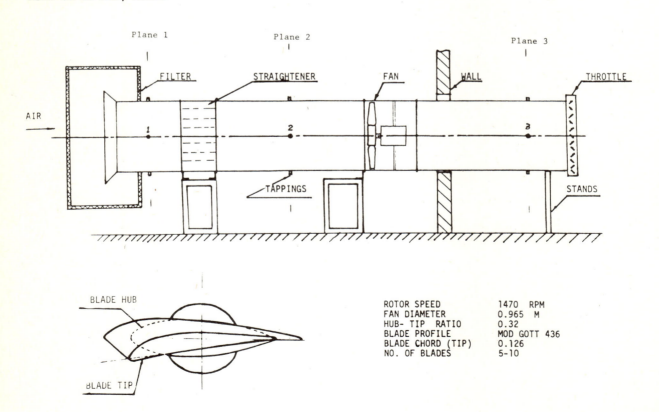

Fig 1    Low-speed fan test facility

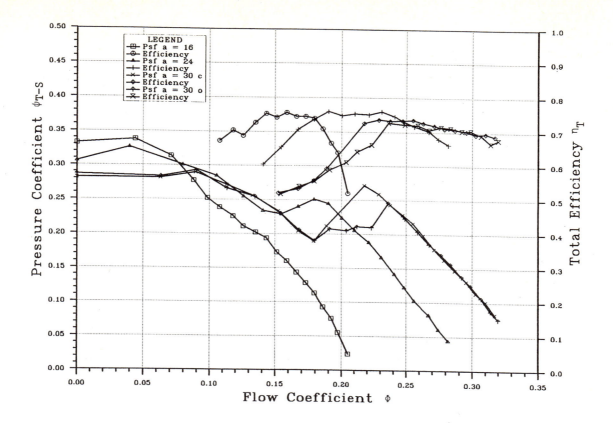

Fig 2    Fan static pressure characteristics (build 1: Gott 10; speed: 1470 r/min; pitch angle: 16°, 24°, 30°)

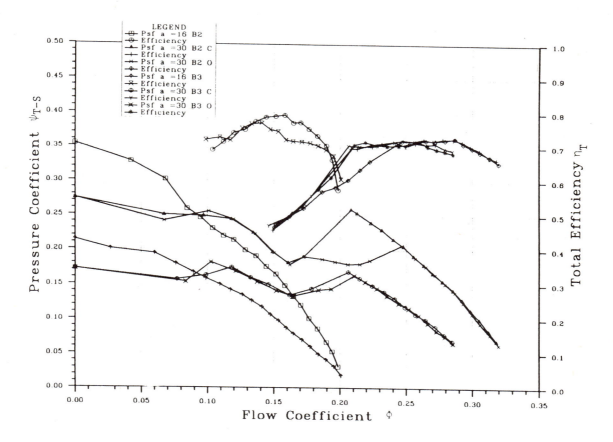

Fig 3    Fan static pressure characteristics (speed: 1470 r/min; build 2: Gott 10 + OGV; build 3: Gott 5)

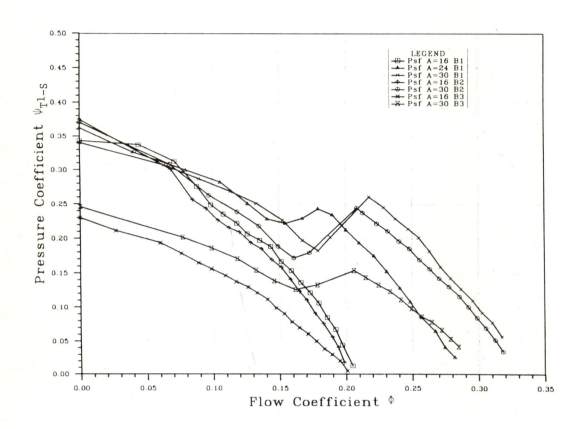

Fig 4     Fan total static pressure characteristics

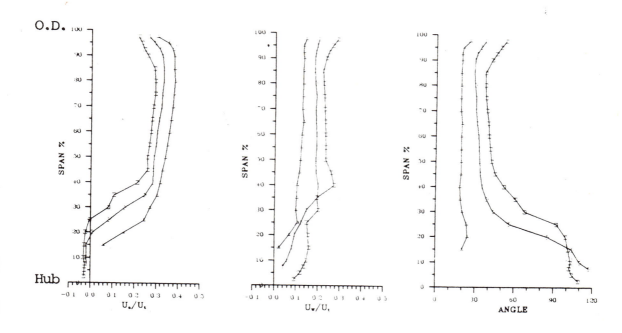

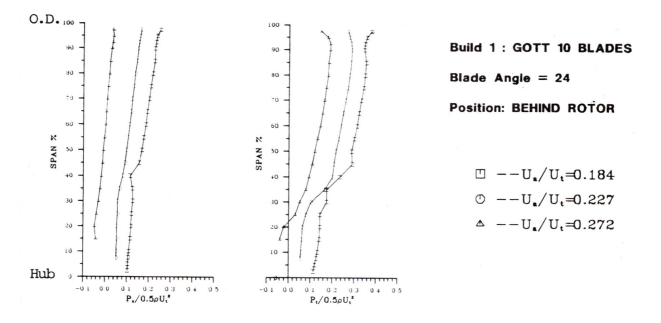

**Build 1 : GOTT 10 BLADES**

**Blade Angle = 24**

**Position: BEHIND ROTOR**

☐ $--U_a/U_t$=0.184

⊙ $--U_a/U_t$=0.227

△ $--U_a/U_t$=0.272

Fig 5    Radial traverse measurements (build 1: Gott 10 blades; blade angle: 24°;
position: behind rotor)

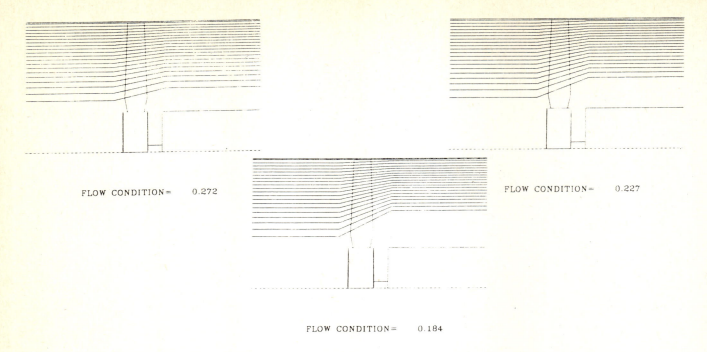

FLOW CONDITION= 0.272

FLOW CONDITION= 0.227

FLOW CONDITION= 0.184

Fig 6     Sketch of the stream lines (build 1: Gott 10 blades, angle: 24°)

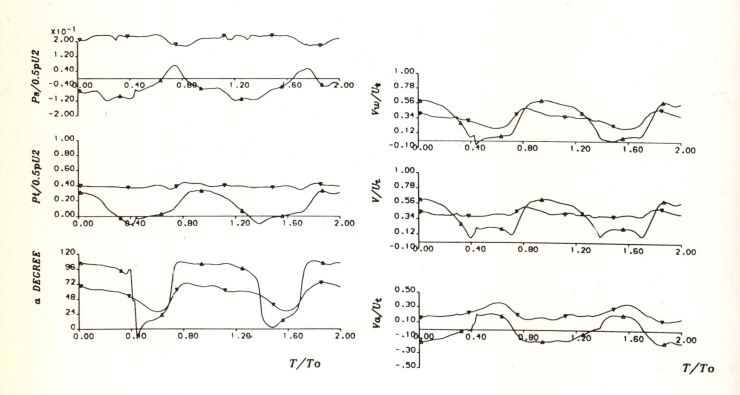

Fig 7     Flow parameter variations at 90 per cent of the span

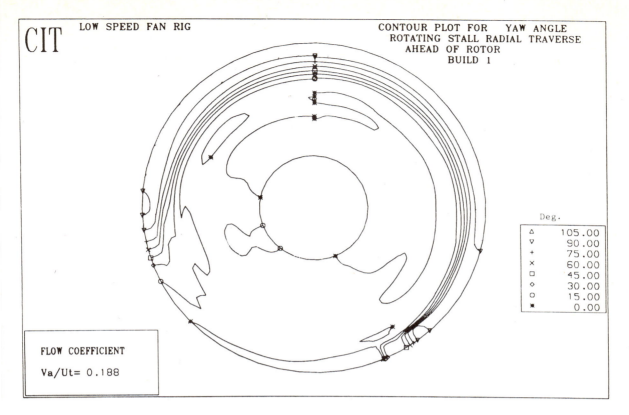

Fig 8a    Contour of yaw angle

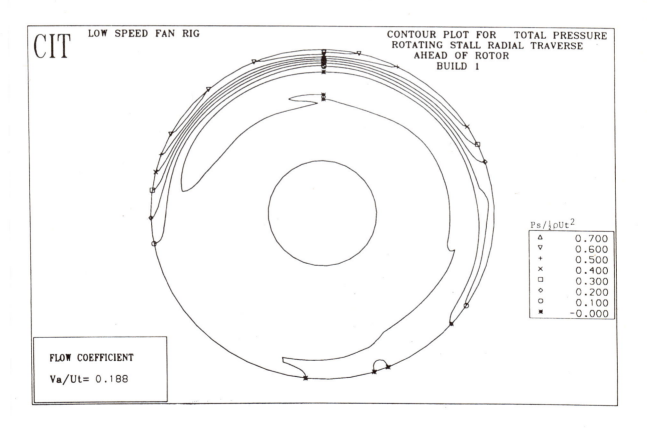

Fig 8b    Contour of total pressure

# C401/012

# Losses of faired struts

**A B McKENZIE**, BSc, DTech, CEng, FIMechE
School of Mechanical Engineering, Cranfield Institute of Technology, Cranfield, Bedford

SYNOPSIS  The significant reduction of duct pressure loss obtained by aerofoil section struts are demonstrated. The losses are broken down into duct friction, strut wake, and interference drag. Comments on noise, incidence, and duct velocity are made.

## 1    INTRODUCTION

The pressure loss due to struts and other obstructions in fan ducts can be considerable if attention is not given to their aerodynamic design.  This was made very evident by the losses incurred in the by-pass duct of an early turbo fan aero engine.  As one part of investigations into minimising the duct losses tests were made of the effect of various shapes of strut fairings.

## 2    TEST ARRANGEMENTS

Because the by-pass duct, which is an annular passage surrounding the core engine, had a relatively small annular height compared to its circumference, the tests were carried out in a 2" x 10" rectangular duct with the strut spanning the 2" dimension at the centre of the 10" dimension.  Fig 1 is a diagram of the test arrangement, and Fig 2 shows the profile shapes of the struts tested.

The basic strut was a 1" diameter cylinder.  The 1" x 2" maximum cross sectional area was retained for all struts and various fairing shapes were applied.  Since the object was to adopt the simplest and lightest form of fairing the initial two were straight taper tails of differing lengths giving maximum thickness/chord ratios of 33% and 25%.  These were followed by three C4 aerofoils of 20%, 25% and 33% thickness/chord ratio.  A tail fairing using the C4 thickness distribution from maximum thickness to trailing edge giving a thickness/chord ratio of 25% followed.  Finally an elliptical section of 33% thickness/chord ratio completed the series.

## 3    TEST PROCEDURE

The airflow was varied over the range 0.2 to 0.5 Mach No in the inlet section of the duct and the static pressure drop between planes 1 and 2 (Fig 1) was measured with no strut fitted.  Each strut was then fitted in turn and the pressure drop was again measured over the same range of Mach No.

A traverse of total pressure was made at 0.4 Mach No of the 10" duct dimension at the mid span of the strut in plane 2 to determine the strut wake.

## 4.    ANALYSIS OF LOSSES

The losses can most simply be expressed as a total pressure loss coefficient $\Delta P/q$ where q is the upstream dynamic pressure ($\frac{1}{2}\rho V_1^2$).  Since the mass average duct velocity downstream is equal to the mean upstream velocity the total pressure loss is taken as the drop in static pressure from plane 1 to plane 2.

While this form of loss coefficient is useful as a comparative figure for the losses of the various strut sections it is particular to the strut fitted in the test duct.  A more general way of comparing the struts is by expressing the losses as a drag coefficient, $C_D$, which is derived from:-

$$D = C_D \tfrac{1}{2}\rho V^2 .x.y$$

where D is the drag force on a body moving relative to a fluid, $\rho$ is the fluid density, and V the free stream velocity, while x and y are representative linear dimensions of the body.

In the tests concerned the total force in the direction of the flow is given by

$$D = \Delta p.A$$

where A is the cross sectional area of the duct.  This will include the viscous force due to the duct walls, as well as the drag of the strut.  If the pressure drop measured for the empty tunnel is subtracted from the pressure drop measured with a strut in position then the result will be a measure of the drag of the strut.  However, this will still include the drag, or pressure drop, caused by the secondary flows created at the intersection of the strut and the duct walls.  This is often referred to as interference drag.

If a mean total pressure loss is calculated from the wake traverse and this is assumed to apply to the complete span of the strut then a loss coefficient can be calculated which will be indicative of the profile drag only of the strut.

Let $\Delta p_T$ = pressure loss of the duct with strut

$\Delta p_d$ = pressure loss of the empty duct

$\Delta p_w$ = pressure loss due to strut wake

$\Delta p_i$ = pressure loss due to interference effect of strut and duct wall

Then $\Delta p_T = \Delta p_d + \Delta p_w + \Delta p_i$

Thus three drag coefficients can be defined:-

$$C_{D_d} = \frac{\Delta p_d}{\frac{1}{2}\rho V^2} \cdot \frac{A}{2(s+b)l}$$

$$C_{D_w} = \frac{\Delta p_w}{\frac{1}{2}\rho V^2} \cdot \frac{A}{t.s}$$

$$C_{D_i} = \frac{\Delta p_T - \Delta p_w - \Delta p_d}{\frac{1}{2}\rho V^2} \cdot \frac{A}{t^2}$$

It must be noted that each of these drag coefficients is related to a different area. For the strut wake the maximum thickness t and span s obviously provide the appropriate area to be proportional to the wake drag. For the duct wall friction the drag will be proportional to the wall surface area which is given by $2(s+b)l$ where s and b are the cross sectional dimensions of the rectangular duct and l is the length concerned. The interference drag is obviously independent of the span and will be primarily dependant on the strut thickness. An artificial area $t^2$ is therefore used in the drag expression, as recommended by Hoerner (1).

The total drag of any particular strut and duct combination can be calculated as follows:-

$$D_d = C_{D_d} . \tfrac{1}{2}\rho V^2 \, 2(s+b)l = \Delta p_d . A$$

$$D_w = C_{D_w} \tfrac{1}{2}\rho V^2 . t.s = \Delta p_w A$$

$$D_i = C_{D_i} \tfrac{1}{2}\rho V^2 t^2 = \Delta p_i A$$

$$D_t = D_d + D_w + D_i = \Delta p_T . A$$

Hence

$$\Delta p_T = (D_d + D_w + D_i)/A$$

$$\frac{\Delta p_T}{\frac{1}{2}\rho V^2} = (C_{D_d} 2(s+b)l + C_{D_w} t.s + C_{D_i} t^2)/A$$

## 5 TEST RESULTS

### 5.1 Total Loss

The total losses as measured by direct pressure drop are plotted against duct Mach No on Fig 3. These show that the empty duct loss is large in comparison to the loss of an aerofoil strut but comparitively small compared to a cylinder, or poorly faired strut. It also illustrates that up to 0.5 Mach No there is little increase in the loss coefficient of an aerofoil strut but a considerable increase for the cylinder and straight taper tail fairings. This is understandable in that the maximum Mach No around the aerofoil will be comparatively small, whereas for the cylinder or taper tail the maximum local Mach No will approach unity at an upstream Mach No of 0.5. Thus not only is the low Mach No loss improved by the aerofoil but the loss coefficient does not rise significantly in the Mach No range considered. This can be important where, even if the mass average approach Mach No is low, if there is significant maldistribution of the flow (for example, immediately downstream of a bend), the local approach Mach No may be considerably greater.

### 5.2 Profile Loss

The wake measurements of four of the struts tested are shown in Fig 4. These show the dramatic reduction of the wake profile achieved by the progressive change from the cylinder to a 25% t/c C4 aerofoil. The results for all the profiles are tabulated in table I and illustrated as profile drag coefficients in Fig 5. The values for all the C4 aerofoils and the elliptical fairing are 5% or less of the cylinder. It is doubtful whether the experimental accuracy was sufficient to distinguish positively between those which are less than 5% of the cylinder profile drag. It is clear however that the simple tapered tails are inferior to an aerofoil profile even if significantly better than the cylinder.

Table 1 Loss and drag coefficients at 0.4 Mach no.

| FAIRING | t/c | $\frac{\Delta p_{o.a}}{q}$ | $\frac{\Delta p_{w+i}}{q}$ | $\frac{\Delta p_w}{q}$ | $\frac{\Delta p_i}{q}$ | $C_{D_w}$ | $C_{D_i}$ |
|---|---|---|---|---|---|---|---|
| cylinder | 1.0 | 0.316 | 0.236 | 0.186 | 0.050 | 1.86 | 0.25 |
| short tail | 0.33 | 0.219 | 0.139 | 0.070 | 0.069 | 0.7 | 0.345 |
| long tail | 0.25 | 0.111 | 0.031 | 0.014 | 0.017 | 0.14 | 0.085 |
| short C4 | 0.33 | 0.093 | 0.013 | 0.009 | 0.004 | 0.09 | 0.02 |
| medium C4 | 0.25 | 0.089 | 0.009 | 0.007 | 0.002 | 0.07 | 0.01 |
| long C4 | 0.20 | 0.108 | 0.028 | 0.008 | 0.020 | 0.08 | 0.1 |
| C4 tail | 0.25 | 0.102 | 0.022 | 0.008 | 0.014 | 0.08 | 0.07 |
| ellipse | 0.33 | 0.099 | 0.019 | 0.007 | 0.012 | 0.07 | 0.06 |

Duct loss $\frac{\Delta p_d}{q} = 0.08$ : $C_{D_d} = 0.005$

### 5.3 Loss Breakdown

The total loss coefficient, $\Delta p_T/q$, and its components $\Delta p_d/q$, $\Delta p_w/q$ and $\Delta p_i/q$ are given in Table I at a Mach No of 0.4. Using the relationships given previously the wake, or profile drag coefficient $C_{D_w}$ and the interference drag coefficient $C_{D_i}$ are also tabulated and these are shown as a bar diagram in Fig 5.

As indicated previously it is doubtful whether the experimental accuracy was sufficient to distinguish accurately between the best of the fairing profiles. In particular Fig 5 may indicate too high a $C_{D_i}$ for the 20% t/c C4 since this is the only case where $C_{D_i}$ is greater than $C_{D_w}$ and the increase of $C_{D_i}$ from 25% t/c to 20% t/c appears excessive. Since $C_{D_i}$ is obtained as the difference of two larger quantities some inaccuracy is not unexpected.

## 5.4 Reynolds No

All the drag coefficients referred to will vary with Reynolds No. Based on the strut thickness (ie cylinder diameter) the Reynolds No for all struts at 0.4 Mach No was $2.2 \times 10^5$. This is just above the generally quoted critical value for a long cylinder of $2 \times 10^5$. However the variation of loss with Mach No shown on Fig 3 would not indicate any major variation due to Reynolds No immediately above or below the 0.4 Mach No test condition, since Reynolds No is approximately proportional to Mach No for these tests.

## 6 INCREASE OF DUCT AREA

While the use of suitable fairings can effect significant reductions of pressure loss it should also be recognised that an increase in duct cross sectional area can produce equally significant improvements.

Consider a strut of constant thickness spanning the diameter of a circular duct.

$$\Delta p = C_D \tfrac{1}{2} \rho V^2 t.D / \tfrac{1}{4} \Pi D^2$$

$$V^2 \propto 1/A^2 \propto 1/D^4$$

Hence
$$\Delta p \propto 1/D^5$$

If for structural reasons the ratio t/D requires to be kept constant then

$$\Delta p \propto 1/D^4$$

Thus the pressure loss of a constant thickness strut can be halved by a 15% increase of duct diameter or a 20% increase if strut t/D is held constant.

## 7 EFFECT OF INCIDENCE

All the tests were made at zero incidence. It can be the case that a strut in a fan duct is subjected to an incidence angle which may not be known accurately. Clearly a cylindrical strut will give no variation of loss. For a C4 aerofoil the loss would still be much less than for a cylinder so long as the aerofoil is not stalled. Stalling would be expected at an incidence in the region of $12°$ and the loss would rise rapidly at higher angles. Where the incident angle of the flow is known the strut should be arranged to suit. However, if it is not known but could be $15°$ or more it may be best to use a cylinder in the largest practical duct area to minimise the velocity and so the losses.

## 8 NOISE EFFECTS

The larger and more turbulent wake generated by an unstreamlined strut will in itself generate more noise. This will be comparatively small, however, compared to the noise generated if the wake from an upstream strut passes into a fan rotor before it has attenuated. As illustrated by the wake traverses shown in Fig 4 the aerofoil profile struts give a very much smaller wake which will attenuate in a much shorter streamwise distance than that of a non faired strut. Noise reduction is, therefore, another important reason for fairing struts, particularly if they must be a short distance upstream of a fan rotor.

## 9 CONCLUSION

The experiments have shown that by providing streamline fairings to struts across ductwork the pressure losses incurred can be greatly reduced. The advantages of keeping duct velocities low by using the largest practical cross sectional area are also important in the interests of minimising the pressure losses of struts.

## 10 ACKNOWLEDGEMENT

The experimental results were obtained under the author's supervision at Rolls Royce, Derby, whose permission to use them is gratefully acknowledged.

Acknowledgement is also made to Mr A Williamson who originally analysed and reported the work internally at Rolls Royce.

REFERENCES

1. Hoerner S F  –  Fluid Dynamic Drag
Published by the author 1965

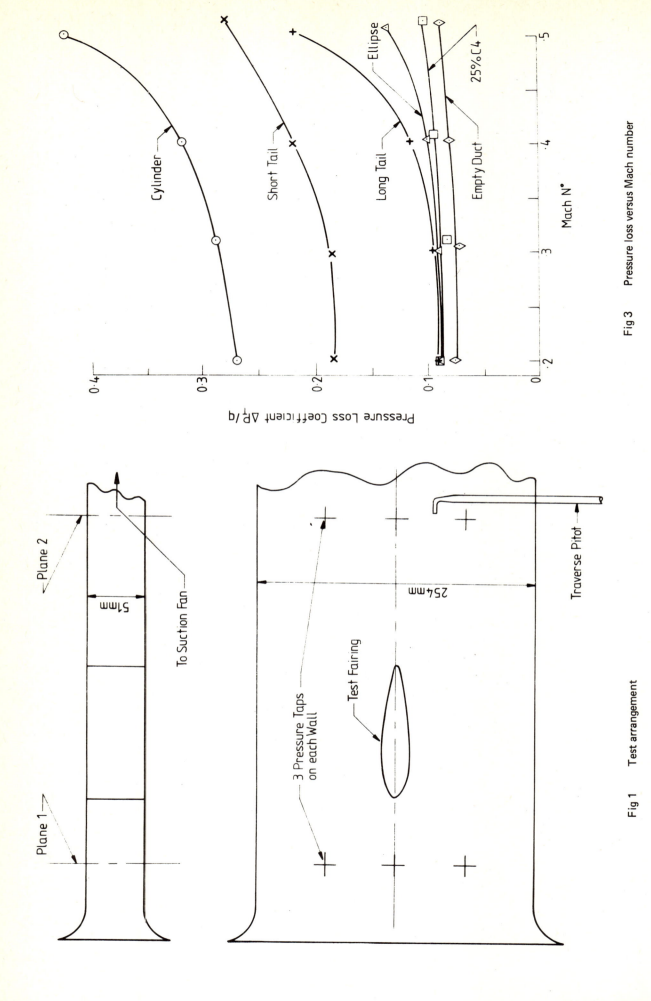

Fig 3    Pressure loss versus Mach number

Fig 1    Test arrangement

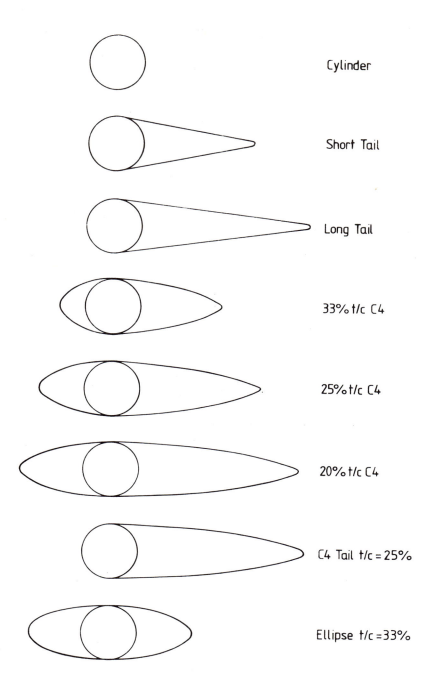

Cylinder

Short Tail

Long Tail

33% t/c C4

25% t/c C4

20% t/c C4

C4 Tail t/c = 25%

Ellipse t/c = 33%

Fig 2 Fairing profiles tested

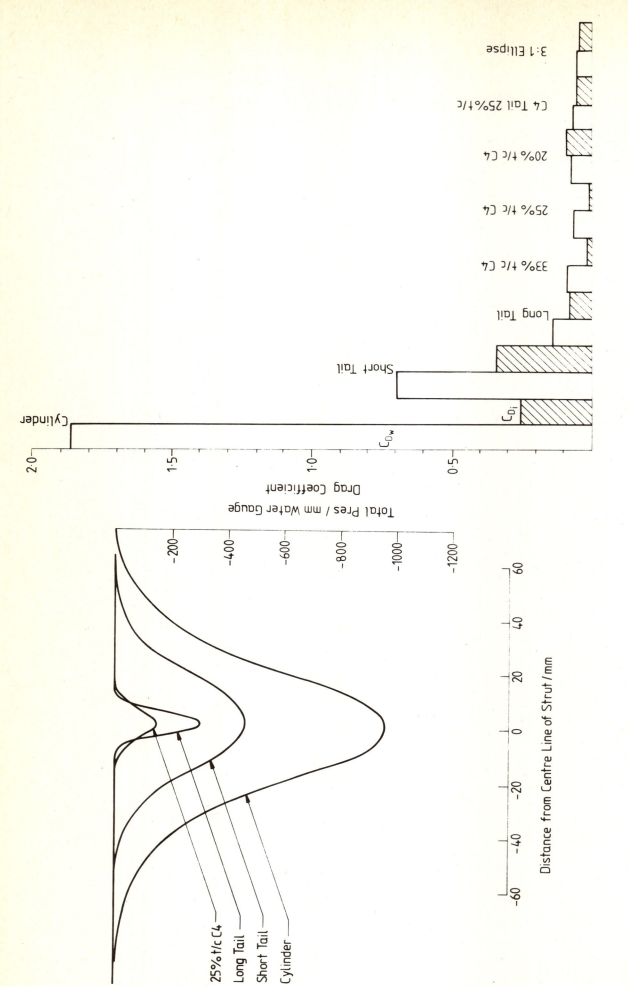

Fig 5    Comparison of wake and interference drag coefficients

Fig 4    Wake traverses at Mach number 0.4

# C401/023

# Effect of outlet condition on small fan performance

**B P MOSS**, OBE, BSc, CEng, FCIBSE, MASHRAE
Nu-aire Limited, London

SYNOPSIS

The way in which very small fans are built into enclosures is critical to their performance. Although this statement applies to all fans, the sensitivity of small impellers often surprises those accustomed to working with large machines. It is an unfortunate characteristic because small fans are often crammed inside crowded enclosures to cool equipment or squeezed within domestic products designed to be attractive rather than encourage aerodynamic efficiency. This paper describes the redesign of a small, domestic, stand-by fan module where modifying the outlet conditions has produced a significant increase in airflow coupled with a reduction in noise.

NOTATION

Psf  fan static pressure, Pa
qv   volume flow rate, l/s
Lw   sound power level, dB$^{-12}$W
LpA  average A weighted sound
     pressure level at 3m distance

INTRODUCTION

The way in which very small fans are built into enclosures and systems is critical to their performance. Poor entry conditions significantly reduce airflow and exit obstructions lead to further reductions in aerodynamic performance, causing turbulence and noise.

These effects are common to all fans but the sensitivity of small fans, especially to their enclosures, often surprises those more accustomed to working with larger machines. Small, forward curved, centrifugal fans, driven by inefficient shaded pole motors, can be particularly difficult to handle because the non linear relationship of their speed/resistance characteristic makes it impossible to forecast their wildly fluctuating flowrates, let alone their noise levels, with any degree of confidence.

This is unfortunate because these small fans are often used in applications where space is severely restricted; crammed into crowded enclosures to cool equipment or squeezed within domestic products designed more to be attractive than to encourage aerodynamic efficiency. And inevitably, noise is a major consideration in these applications.

However, careful attention to detail, when supported by an awareness of the principles involved, can achieve far better results than are normally accepted. This paper describes the redesign and testing of a small stand-by fan module used in a domestic extractor unit where modifying the outlet conditions to the fans produced a significant increase in airflow and a reduction in noise.

The unit in question was conceived some 25 years ago to provide a convenient way of ventilating internal, ie. windowless, WC compartments and bathrooms. It uses standard 100mm internal diameter pipe to carry the extracted air to the outside of the building; much easier to accommodate than the 225mm x 225mm ductwork employed previously. The introduction of the original unit enabled architects to design narrow frontage dwellings with a greater degree of freedom than had been possible using the bulky equipment available previously. As a consequence, a large number of manufacturers throughout Europe have been encouraged to produce similar equipment and many tens of thousands of units are now sold each year.

Because the original brief called for as small a unit as possible to extract through a 100mm i.d. pipe; and for a pressure characteristic which would enable the unit to push air along relatively long and, often, convoluted, pipes; a small forward curved multivane impeller was selected as the most appropriate fan. This feature has not changed and the forward curved multivane is used, almost universally, by all manufacturers.

A significant proportion of the original building development took place under, what was then, GLC regulation and, therefore, the unit was produced, almost immediately, in a twinfan version which offered a stand-by facility should the running fan fail. Airflow sensing and automatic fan changeover were introduced at the same time and subsequently an astonishing variety of special controls have been introduced in response to customer demand. This paper will concern itself to discussion of the twinfan, standby, models only.

The original unit was housed in a simple, all metal, casing typical of enclosures used for low volume production. It is now housed in an injection moulded casing, more suited to the higher sales associated with this type of product. The new unit housing will be built-up from a selection of basic components produced using a multiform tool, enabling models suitable for surface, flush and duct installation to be easily assembled.

From its inception, the range has always been produced to meet the high duty specification of the "engineer-selected" market whereas seemingly similar but lower performance equipment has been directed more at the distributor and domestic sectors. This accounts for the somewhat larger size, better aerodynamic performance and quietness of the range when compared to others on the market.

However, over the years, various modifications, cumulatively, had reduced the overall performance of the equipment and this, together with a perceived market demand for increasing airflow rates and less noise, made necessary a radical evaluation of the design.

## 2 CURRENT DESIGN

Figure 1 shows the general arrangement of the outlet configuration, The rectangular outlets from two small, 72mm diameter, multivane impellers are connected directly to a 100mm i.d. spigot. Quite clearly, the airflow discharging from the working fan is subject to violent obstruction as the cross-sectional area of the passage is reduced, momentarily, by, some, 20 per cent before expanding immediately into the 100mm i.d. outlet section.

Subsequently, the airflow has to negotiate a spring-loaded butterfly shutter before entering the outlet pipe itself. This element causes further turbulence

## 3 PROPOSED DESIGN

### 3.1 Design One
Figure 2 shows the initial proposal for improving the outlet condition. The two rectangular outlets are connected to the 100mm i.d. section via a 125mm i.d. transformation piece. This guides the airflow into the outlet pipe more cleanly and without reducing the passage cross-sectional area to less than that of the fan outlet itself. The backdraft

shutter is unchanged.

A model was constructed and tested. The airflow rate increased by, some, 20 per cent but the design was rejected on the grounds that the 125mm transformation section would have prevented, or made unacceptably difficult, substitution of an existing unit by a new type replacement.

### 3.2 Design Two
Figure 3 shows the second proposal for improving the outlet condition. The two fan scroll outlets are now combined into a single rectangle which fits more easily into the 100mm i.d. outlet section, eliminating the need for the transformation section. This is achieved by merging the two scrolls into a composite unit where they share a common outlet and using an air operated shutter to seal off the stand-by fan scroll.

Although this design guides the airflow into the outlet pipe cleanly and without reducing the passage cross-sectional area to less than that of the fan outlet itself, it was realised that the necessity to retain a backdraft shutter in the outlet spigot would cause turbulence and, hence, noise and resistance and the proposal was rejected.

### 3.3 Design Three
Figure 4 shows the third proposal for improving the outlet condition. As was the case with Design Two, the two scrolls are combined into a single outlet, however, the simple changeover shutter suggested for the second design is now replaced by a mechanism which combines the functions of both changeover and back draft shutters.

This mechanism consists of two separate and identical shutters, independently pivoted at the merging of the fan scrolls. When one of the fans is running they move as one and act as a simple changeover shutter preventing "blow back" through the stand-by fan and they move, again together, to open the stand-by fan scroll outlet and seal the "duty fan" scroll outlet should the "duty fan" fail. However, the two shutters also incorporate magnets close to their hinge axis and these are orientated to repel the shutters. Thus, when neither fan is running, the two shutters move apart and seal both scroll outlets.

It was found necessary to incorporate small "dams" on the swinging edge of the shutter leaves as, otherwise, there was a tendency for them to be "sucked" by the impeller rotation into the running fan scroll. An alternative solution to this problem was to reduce the area of the scroll outlets to increase the velocity of the airflow. However, the dams have proved effective, allowing the optimum design of scroll to be used.

## 3.4 Design Four

The scroll configuration and shutter mechanism of Design Three were approved and more detailed attention was now given to the method of sensing "fan failure".

Plastic flags, connected by stiff wires to externally mounted rotary micro-switches, are deployed in the scroll passages of current units to detect whether or not acceptable exhaust rates are being achieved. The system was introduced in the earliest stages of development of the range and has proved extremely reliable. However, in spite of considerable refinement, the presence of flags in the airflow reduces the aerodynamic performance and, at the very commencement of the design exercise, it was decided to investigate alternative methods of sensing the duty fan performance. Bearing in mind that introducing anything into the airstream would inevitably cause turbulence, development concentrated on devices located "outside" the scroll/pipe system and the final choice came down in favour of a Hall Device sensing the proximity of small magnets fitted in each of the shutter leaves.

The final design, therefore, consists of the scroll/shutter configuration of Design Three combined with a Hall Device airflow sensing system.

The arrangement is the subject of a patent application.

## 3.5 Design Five

The multiform plastic injection moulding tool enables the casings of all model types: surface, recessed and duct, to be increased in depth by adding sections and this feature led to the development of further models, incorporating larger fans, and intended for applications which demand higher ventilation rates. All the design principles of Design Four are included in the higher duty models.

Because the higher duty range is entirely new to the Company, it is not possible to compare its performance with an existing "made in house" model. However, it can be compared with competitive units which use identical fans. The result of this exercise is displayed, graphically, in the Results Section of this paper.

## 4   RESULTS

The following graphs and tables display aerodynamic and acoustic values measured in a test house approved by the British Standards Institution as being an integral part of a facility qualified to BS5750:Part 1 and further inspected, supervised and approved to carry out testing for the Certification of Air Moving Equipment Scheme.

... All testing was carried out in accordance with the latest edition of BS848; aerodynamic measurement to Part 1, Inlet Chamber Method and acoustic measurement to Part 2, Reverberation Room Method.

In order to simplify the presentation of data but, at the same time, maintain valid comparisons, all the aerodynamic and acoustic measurements used in this paper were obtained with the equipment under test mounted on a vertical surface. Tests confirmed that similar comparisons are obtained when the equipment is installed in other planes.

Sound Levels were measured on the inlet side of the fans and all acoustic tests were performed with the units operating against a similar, nominal, resistance.

Because the purpose of this paper is to demonstrate only how altering the outlet condition changes the performance, any effects which may have been caused by the current and new enclosures presenting dissimilar aerodynamic and acoustic resistances were eliminated by performing all testing with the covers removed.

For a similar reason, flag switches were not fitted.

A disadvantage of testing the units unenclosed and without flag switches is that the aerodynamic and acoustic information so obtained is not, in absolute terms, the performance of a complete unit. However, it does ensure that the results meaningfully and accurately demonstrate the effect of changing the fan outlet condition and nothing else.

Rather than use scaling methods to predict what would be the noise level of the new unit if it were extracting at the same rate as the current unit, sound levels were measured directly when the performance of the new equipment was made to match that of the current unit by adjustment of its fan speed.

Table 1  Comparison of acoustic performance of current
and new units.

| frequency Hz | Lw 63 | 125 | 250 | 500 | 1k | 2k | 4k | 8k | LpA |
|---|---|---|---|---|---|---|---|---|---|
| current unit | 46* | 52 | 50 | 48 | 50 | 46 | 39 | 31 | 36 |
| new unit | 48* | 48^ | 47 | 46 | 46 | 41 | 36^ | 29* | 32 |

\* Difference between measured sound level and background
  level less than 6dB.  Highly suspect.
^ Difference between measured sound level and background
  level between 6dB and 10dB.  Suspect but may be quoted.

Table 2  Comparison of acoustic performance of
current and new units with the aerodynamic performance of
the new unit reduced to that of the current unit.

| frequency Hz | Lw 63 | 125 | 250 | 500 | 1k | 2k | 4k | 8k | LpA |
|---|---|---|---|---|---|---|---|---|---|
| current unit | 46* | 52 | 50 | 48 | 50 | 46 | 39 | 31 | 36 |
| new unit | 46* | 47* | 47^ | 43 | 42 | 35^ | 30* | 26* | 28 |

\* Difference between measured sound level and background
  level less than 6dB.  Highly suspect.
^ Difference between measured sound level and background
  level between 6dB and 10dB.  Suspect but may be quoted.

## CONCLUSIONS

The new standard range extracts, some,
35% more air than does the equipment it
is replacing.  Fig.5, which includes a
typical system curve, clearly
demonstrates this improvement.

In spite of the increased airflow,
the new units are also, some, 4dBA
quieter, a significant reduction.

If noise is the major
consideration, or the present extract
rate is acceptable, the new range is,
some, 8dBA quieter than the current
equipment, representing almost a
"halving" of the perceived noise.

The aerodynamic performance of the
higher duty fan unit is substantially
above that produced by the identical
impeller/motor assembly when fitted in
competitive equipment which is selling
successfully in the marketplace.

Modifying the outlet conditions has
resulted in very significant
improvements in the performance of the
new equipment.  The design of the fan
outlet configuration was concerned with
small details but this was to be
expected bearing in mind the scale of
the equipment itself; nevertheless,
clearly it was based on an awareness of
the aerodynamic principles involved.
The exercise demonstrates that paying
meticulous attention to the way in which
small fans are used is a worthwhile
investment if aerodynamic and acoustic
performance are market criteria.

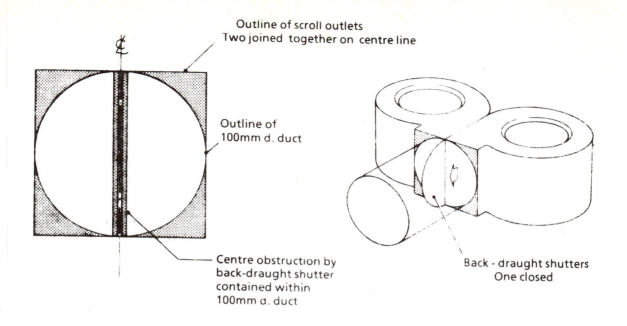

Outline of scroll outlets
Two joined together on centre line

Outline of
100mm d. duct

Centre obstruction by
back-draught shutter
contained within
100mm d. duct

Back - draught shutters
One closed

Fig 1      Outlet condition — current design

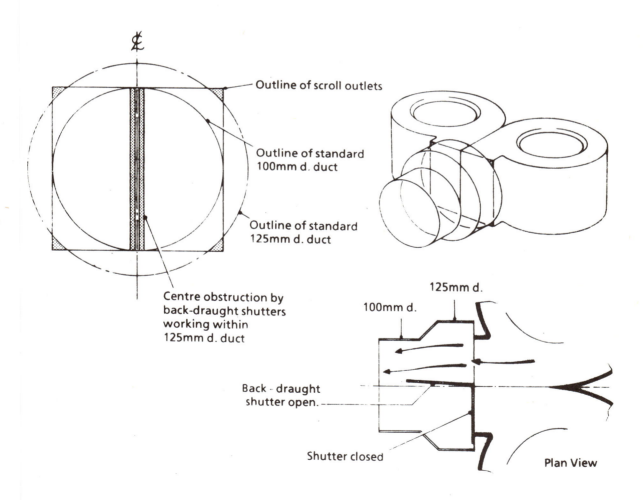

Outline of scroll outlets

Outline of standard
100mm d. duct

Outline of standard
125mm d. duct

Centre obstruction by
back-draught shutters
working within
125mm d. duct

125mm d.

100mm d.

Back - draught
shutter open.

Shutter closed

Plan View

Fig 2      Outlet condition — initial proposal for redesign

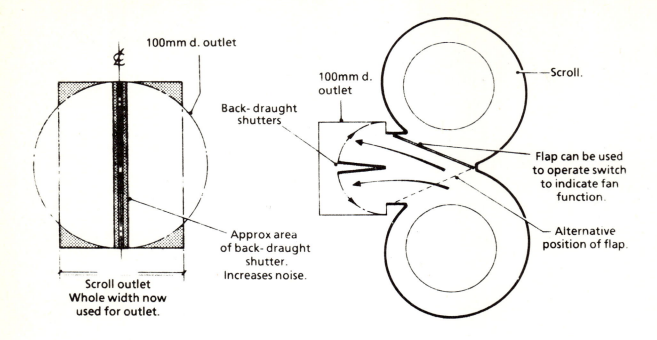

**Fig 3    Outlet condition — second proposal for redesign**

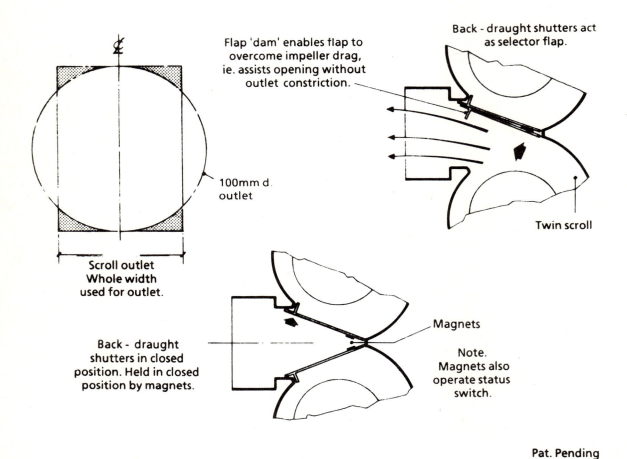

**Fig 4    Outlet condition — final redesign**

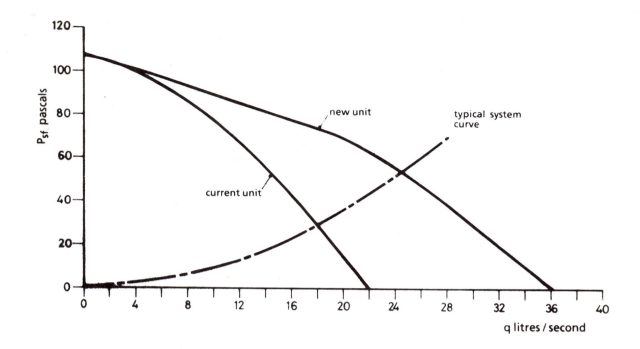

Fig 5    Aerodynamic performance. New unit versus current unit

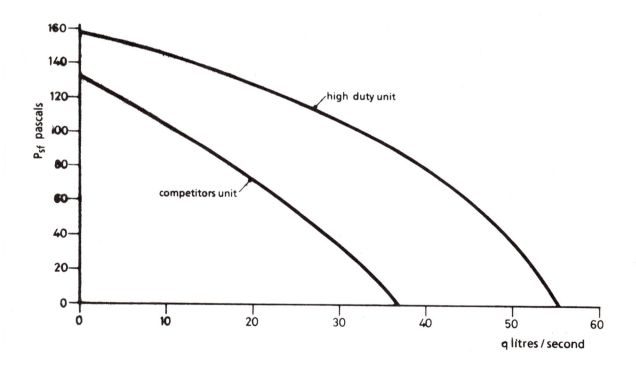

Fig 6    Aerodynamic performance. High duty unit versus competitors unit

# The development of axial flow fans for the venting of hot fire smoke

**G A C COURTIER**, DFH, CEng, FIEE and **J A WILD**, CEng, MIMechE
Woods of Colchester Limited, Trafford Park, Manchester

The paper outlines the development of Axial Flow Fans for the venting of hot fire smoke, from an investigation into the technology of calculating ventilation rates to testing and certification.

## 1    INTRODUCTION

This paper deals with an installation effect that in most cases will never arise, but nevertheless is assuming increasing importance in fan system design, that is, the response of the system to the emergency conditions that arise in the event of a fire.

While at one time the natural reaction was that a fire is best dealt with by depriving it of oxygen, now the maintenance of air flow is considered essential, for two basic reasons:

(a)   To keep escape routes clear of smoke and thereby facilitate the safe evacuation of occupants.

(b)   To assist the Fire Brigade -

by enabling them to find the seat of the fire on arrival at the site.

by preventing or limiting the phenomena of either flashovers which can occur when the smoke temperature reaches 600°C, or backdraughts, which result from a build-up of carbon monoxide in the hot smoke.

### 1.1   History of Smoke Venting

Early smoke venting systems were fitted into single storey factory-type buildings. Distances the smoke would travel were usually short and hence the smoke was assumed to have retained sufficient buoyancy to operate the roof mounted natural vents provided.

The demand for fans to ventilate hot fire smoke began to accelerate during the early 1980's Three possible reasons are thought to be responsible for this change:

(a)   Smoke venting was now being provided to more complex building types often requiring ductwork to convey the hot smoke to the outside of the building.

(b)   The increasing use of the existing ventilation system exhaust fans to cover the dual role of providing the smoke venting cover, either single or two speed.

(c)   The growing knowledge which was indicating that the temperature, and hence the buoyancy of the smoke, may not be as high as had been earlier supposed. This in turn would limit the effectiveness of the traditional natural vent - particularly during the early stages of the fire when people are escaping.

### 1.2   The Temperature/Time Specification

In addition to their normal requirement of moving a quantity of gas against a given resistance fans used for venting hot fire smoke must be able to handle the hot smoke, at a given <u>temperature</u> for a pre-determined <u>time</u> period.

This has become known as the Temperature/Time Specification, or fire rating.

The results of a survey of current Temperature/Time Specifications from various countries, is detailed in Table 1:

Temperature/Time Specifications by country

| Country | Temp °C | Time Hrs. | Comments |
|---|---|---|---|
| Australia | 200 | 2 | |
| Canada | – | – | Each consultant legally responsible for safety |
| Hong Kong | 250 | 1 | |
| Italy | 200<br>400 | 2 or 3<br>2 | |
| Belgium | 250<br>400 | 2 or 3<br>2 | |
| France | 200<br>400 | 2<br>2 | Certificate of Independent Test required by law |
| W Germany | 600 | 1.5 | |
| Malaysia | 250 | 2 or 3 | |
| New Zealand | 300 | 0.5 or 1 | Following UK practice |
| Singapore | 150<br>250 | 1<br>1 | Likely to adopt UK practice |
| Egypt | 400 | 2 | French consultant specifications adopted |
| Finland | 350 | 1 | |
| Saudi Arabia | – | – | No current Specification |

Table 1 cont'd

| Country | Temp °C | Time Hrs | Comments |
|---|---|---|---|
| South Africa | - | - | Follow UK, USA |
| U.K. | 300 | 0.5 or 1 | |
| U.S.A. | 650 | 1 | No National app- |
| | 260 | 1 | roach: State Reg- ulations vary |

This survey, augmented by specific enquir-
ies, showed that temperatures varying between
150°C and 650°C were being specified, at time
periods up to 3 hours.

To understand why, it was decided to research fully the technology of fire smoke venting
and thereby establish how ventilation rates and
smoke temperatures from fires are calculated.

This was considered to be an essential preliminary to writing the Specification of a fan
to meet the needs of the application.

## 2   INVESTIGATIONS INTO FIRE SMOKE VENTING TECHNOLOGY

### 2.1   The Temperature Specification

Unlike normal ventilation systems, extraction
rates for fire smoke venting have little to do
with the size of the room, because the amount of
smoke produced depends largely on the size of
the fire.

As the smoke plume rises, surrounding cool
air is entrained into the plume and becomes well
mixed with the hot smoky products of combustion
so as to form an inseparable component of the
smoke.

The quantity (mass) of smoke produced by a
fire will depend on three factors:

(a)  the perimeter of the fire

(b)  the temperature of the flames in the plume

(c)  the effective height of the column of hot
     gases above the fire.

### 2.1.1   Large Fire Theory

The Large Fire Plume Theory, developed at the
Fire Research Station, and outlined in Fire Research Technical Paper No.7(1), established a
relationship between these factors and enables
the mass of smoke produced to be estimated, see
Figure 1.

The mathematical form of this relationship is
stated below:

$$M = 0.096 \, P \rho_0 \, Y^{\frac{3}{2}} \left( g \cdot \frac{T_0}{T} \right)^{\frac{1}{2}} \text{ kg/sec, where}$$

M = mass rate of smoke production   - kg/sec

P = perimeter of fire               - metres

Y = height of smoke column          - metres

$\rho_0$ = density of ambient air               - kg/m³

g = acceleration due to gravity          - m/sec²

$T_0$ = absolute temperature of ambient air   - K

T = absolute temperature of flames       - K

The two most important factors are the perimeter of the fire and the height of the smoke
column.  These two factors affect the smoke production linearly, and to the power of 1.5 respectively.

Since smoke production only varies with the
square root of the absolute temperature of the
fire, this factor is much less important.

If we assume a flame temperature of 800°C,
an ambient average temperature at 17°C (density
1.22kg/m³) the mass of smoke produced can be
obtained by the simple expression:

$$M = 0.19 \, PY^{1.5} \text{ kg/sec } -----(2)$$

A variation of this simple expression
applies when smoke from a fire in a room is
allowed to ventilate into an enclosed area
before being exhausted to atmosphere.  Shops
along a shopping mall are a practical example as
illustrated by Figure 2.

At the point where the smoke leaves the
shop, an additional large amount of air is entrained into it, which affectively doubles its
mass and rapidly cools the smoke.

The simplified expression then becomes:

$$M = 0.38 \, PY^{1.5} \text{ kg/sec}$$

which in practice is often rounded up to read:

$$M = 0.4 \, PY^{1.5} \text{ kg/sec } -----(3)$$

Although other plume theories do exist, the
Large Fire Theory has become generally accepted
and is the one used by designers of smoke venting
systems in the UK.

These expressions enable the designer to
establish the mass rate of smoke production in
kg/sec.

The ventilation rate in (m³/sec) and the
smoke temperature in (°C) can now be calculated
as follows:

$$\Theta = \frac{Q_s}{M} \text{ °C}$$

$\Theta$ = temperature rise of the smoke in (°C)

$Q_s$ = heat carried by the smoke in (kW)

M = mass of smoke produced in (kg/sec)
Specific heat is omitted, being close to unity.
Thus, numerically,

$$V = \frac{M(\Theta + 290)}{354} \text{ m}^3/\text{sec}$$

V = ventilation rate in (m³/sec)

$\Theta$ = temperature rise in (°C)

M = mass of smoke produced in (kg/sec)

## 2.1.2. Design Fire (Sprinklered Areas)

It can be seen from the foregoing that the factors which influence both the ventilation rate and the smoke temperature are:

$Y$ = the height of the smoke column

$Q_s$ = the heat carried by the smoke

$P$ = the perimeter of the fire

The height of the smoke column ($Y$) will always be a variable, determined by the geography of the building. Where people are present, it should not be less than three metres.

The perimeter of the fire ($P$) is a measure of its size and will vary. With sprinkler controlled fires, the spacing of the nozzles is usually taken as a measure of fire size.

Hence a sprinkler grid of 3m x 3m defines the fire size

and $P$ = 12 metres

The heat carried by the smoke ($Q_s$) is estimated from the total heat output of the fire ($Q$) minus heat losses, i.e. $Q_s = Q$ - losses.

$Q$ is the product of the area of the fire in metres and the burning rate of the fuel in kW/m².

i.e. $Q$ = Area (m²) x Burning Rate (kW/m²)

It can readily be seen that $Q$ and $P$ are inter-related.

Some typical burning rates of different materials are given in the following Table:

Table 2 Typical Burning Rates

| Material | Burning Rate kW/m² |
|---|---|
| Stacked chipboard | 85 |
| Books | 93 |
| Furniture | 93 |
| Crated furniture | 100 |
| Cellulosics general | 160 |
| Cardboard reels | 210 |
| Vehicles/petrol/paint | 260 |
| Cartons/electrical goods | 310 |
| Stacked cardboard | 320 |
| Stacked/sawn timber | 390 |
| Cardboard cartons | 630 |
| Ind. Methylated Spirits | 740 |
| Light fuel oil | 1475 |
| Petrol | 1590 |
| Retail store | 500 max |

With this background, a design fire can be specified.

For example - a retail store would have sprinklers on a 3m x 3m grid and sufficient fuel to support a burning rate of 500kW/m² max.

| Fire size | = | 3m x 3m |
|---|---|---|
| Perimeter | = | 12 metres |
| Area | = | 9m² (rounded up to 10m²) |
| Burning rate | = | 500kW/m² |
| Total heat output | = | 5000kW |

The design fire can thus be defined simply as 3m x 3m x 5MW

Table 2 shows that a burning rate of 500kW/m² is, in the main, only exceeded by liquid fires. It can, therefore, be taken as a maximum for the contents of general commercial buildings, and is recommended by the Fire Research Station for retail stores.

## 2.1.3 Design Fires (Unsprinklered Areas)

An insight into the behaviour of fires in unsprinklered buildings is provided by E G Butcher and presented in his paper "Fire Progression, Spread and Growth" (4). Here Butcher calculated how long it would take for a fire, doubling its area every 4 mins, to reach a total heat output of 5MW. This rate of growth is typical of a textile fire, as is regarded as average for all fires. Smoke temperatures are based on a smoke layer height of 4m and an ambient temperature of 17°C. The results are calculated below.

Table 3 Fire Growth

| Time from ignition (mins) | Heat output (MW) | Fire size (m²) | Smoke temperature maximum °C |
|---|---|---|---|
| 0 | 0.1 | 0.375 | 43 |
| 8 | 0.4 | 1.5 | 70 |
| 16 | 1.5 | 6.0 | 118 |
| 20 | 3.0 | 12.0 | 159 |
| 22.5 | 5.0 | 20.0 | 201 |
| 24 | 6.2 | 24.0 | 225 |
| 28 | 13.5 | 48.0 | 337 |

Table 3 shows that an unsprinklered fire could take 22½ minutes to reach a total heat output of 5MW. By that time, its area would be 20m² and its perimeter 18m. The average burning rate over this increased area wuld be only 250kW/m² not the 500kW/m² assumed with the sprinklered fire. As is seen later (paragraph 2.2) a time of 20 minutes is considered a safe figure to allow the evacuation of all people to a safe place, so the design fire of 5MW can be regarded as a valid basis for the fan system design, for both sprinklered and unsprinklered areas.

Hence a sprinkler controlled fire of 5MW would produce higher smoke temperature, because of the smaller perimeter. It was decided to use a design fire of 3m x 3m x 5MW as the basis for writing the temperature specification of the fan range.

Table 4 (overleaf) details the resultant values of smoke volume and temperature at increasing values of $Y$ for both variants and the large fire theory.

| Height of Smoke Layer | Shops etc - ($M = 0.15PY^{1.5}$) | | | Malls - ($M = 0.38PY^{1.5}$) | | |
|---|---|---|---|---|---|---|
| | Mass Rate (kg/sec) | Volume Rate (m³/sec) | Smoke Temp. (°C) | Mass Rate (kg/sec) | Volume Rate (m³/sec) | Smoke Temp. (°C) |
| 2.5 | 9.0 | 22 | 573 | 18.0 | 29 | 295 |
| 3.0 | 12.0 | 24 | 437 | 24.0 | 34 | 227 |
| 3.5 | 15.0 | 27 | 350 | 30.0 | 39 | 184 |
| 4.0 | 18.3 | 29 | 291 | 36.6 | 44 | 154 |
| 5.0 | 25.5 | 35 | 213 | 51.0 | 56 | 115 |
| 6.0 | 33.5 | 42 | 166 | 67.0 | 69 | 92 |
| 8.0 | 51.5 | 57 | 114 | 103.0 | 99 | 66 |
| 12.0 | 95.0 | 92 | 70 | 190.0 | 170 | 43 |

Table 4  Extraction rates and smoke temperature

### 2.1.4 Heat losses from fire

Table 4 assumes that the total heat output of the fire (5MW) is entering the smoke. In practice, this would not be the case as some heat loss would occur.

A fire burning in the open would lose about 25% of the generated heat by radiation.

A fire burning in a room causing hot smoke to issue through an opening, would lose heat to the walls and ceiling. This would vary during the fire but it is probably adequate for design purposes to assume 50% heat loss.

These percentages apply in the absence of sprinklers.

However, it is current convention to ignore these losses and although this practice is being seriously questioned, it was felt that a fan range, with a number of temperature steps, based on the current practice, would adequately cover any potential changes in practice.

### 2.1.5 Temperature element of specification

The foregoing work led to the conclusions that any fan range would need to have a number of temperature/time steps, up to a maximum of 650°C (flashover) to provide the system design engineer with both flexibility and economy. The temperature specification was established as follows:

650°C - This would be the maximum being above flashover when the fire was out of control.

400°C - Motor in airstream fan (axial) - maximum temperature for this fan type.

300°C - Most important step. Would cover over 80% of all designs. Motor in airstream fan.

200°C - Would cover over 60% of all design using standard components.

### 2.2 Time element of specification

Establishing the time element of the temperature/time specification proved somewhat easier.

Current law in the UK enforces a building designer to provide sufficient means of escape

to evacuate people to a safe place in 2½ minutes.

In practice, it is generally accepted within the fire services that most buildings can be cleared within 20 minutes. Hence the general UK specified time period of 30 minutes.

Any longer period would be to assist the fire brigade in either fighting the fire or to clear any residue smoke. Table 1 indicates that views on time will vary in different countries. Within the UK, a time of up to 1 hour is usual for this purpose. Lifetimes of over 2 hours are difficult to justify, unless the fire brigade are delayed by distance or traffic congestion!

Combining this with the temperature steps in paragraph 2.1.5., we can establish the temperature/time specifications for a range of fire smoke venting fans.

These were given category codes and are outlined in the following Table:

Table 5  Temperature/time specifications

| Category Code | Suitable for duties: | | |
|---|---|---|---|
| | Temperature (°C) | | Time (Hrs) |
| 650/1 | 650 | for | 1.0 |
| | 600 | for | 1.5 |
| 400/2 | 400 | for | 2.0 |
| 300/1 | 300 | for | 1.0 |
| 300/0.5 | 300 | for | 0.5 |
| | 250 | for | 2.0 |
| 150/5 | 200 | for | 0.5 |
| | 150 | for | 2.0 |

### 3  FAN SPECIFICATIONS

### 3.1  Fan Requirements for fire smoke venting

The full requirements of a fan range for the venting of hot fire smoke, can be listed as follows:

(a)    To meet the temperature/time specification outlined in Table 5.

(b)    To be able to handle air at normal ambient temperatures so as to

    (i)    Provide for the normal ventilation requirements of the building

    (ii)   Handle the cool smoke present during the early stages of the fire

    (iii)  Facilitate periodic testing

(c)    To have flexibility of duty sufficient to cover the many levels of ventilation rate that calculations based on the large fire theory will demand.

(d)    To be capable of either duct or roof mounting, the latter in a weathering cowl.

## 3.2.  Fan Product Options

It was felt that axial flow fans, rather than centrifugal fans, would best cover these requirements, so the product development and testing programme was concentrated on the axial flow fan.

Next, it was believed to be necessary to adopt some degree of rationalisation when analysing the variety of specifications that are currently encountered, and the temperature/time categories listed in 2.2 are a logical response to this - but clearly not the only possible set.

Until an international consensus enables the variety to be reduced, fan manufacturers will have to determine what options they will offer, based partly on factors of manufacturing costs and partly on their marketing perception.

Nevertheless, any comprehensive range is likely to include one set of products with the drive motor in the airstream and another set where the motor is separated from the main air or gas flow.

Separation from the main flow in the context of axial flow fans will usually be achieved with the bifurcated type of construction, where the motor is in its own compartment. For this to remain cool it also implies that the fan is connected to a ducted system with the plant room remote from the likely heat zone, as distinct from a roof-mounted extract fan which, by its nature, will be located in the fire zone.

With a bifurcated fan operating at very high temperature, even though the drive motor is out of the main airstream, it is important to recognise that heat transfer to the motor cooling air circuit can be substantial:  as an example, a 1250mm fan with 45kW motor discharged 60kW of heat into the plant room region during the 1½ hour test period and means of its removal clearly would need to be allowed for, or a separate ducted source of cool air provided to the intake of the motor compartment. Furthermore, ductwork associated with the fan system needs to be lagged in order to limit temperature rise wherever there is critical wiring or plant in the vicinity.

## 4    INTERACTION BETWEEN DUTY AND DESIGN

Considerations such as these, together with information concerning the short-term thermal properties of materials and structures, provide the "data set" which determine the product design.  The principal aspects of these are:

## 4.1  Structural materials

The decisions for most manufacturers will be at what combination of temperature and stress must aluminium give way to steel, for the impeller, and then at what conditions will conventional mild steels cease to be adequate and heat-resistant steels need to be adopted. At the highest temperatures, the effect of stress-relief of steel casing structures also needs to be considered.

## 4.2  Impeller clearances

Knowing the temperature specified and the materials selected, the dimensional tolerances for running clearances can be defined, and it is in this context that the most obvious effect on fan performance arises, with the well known loss of pressure development and efficiency consequent upon an increase in clearance, and hence leakage and secondary flow (see Figure 3).

It may not always be clear, however, what allowance for differential expansion should be made, since the temperatures expected outside the casing may be unpredictable, especially in the non-steady state conditions that are likely to exist in an emergency.  A safe estimate is to assume an effective casing temperature mid-way between the specified internal temperature and a normal ambient value.

Some fans may only be required to operate in emergency conditions (and during periodical testing), in which case no allowance for increased running clearance need be made.  Most fans for smoke venting though, have a normal duty also, and in this situation, the effect of increased running clearance will affect the fan characteristic and possibly require a different fan selection.

## 4.3  Motor drives

Any bearings in motors or layshafts exposed to the hot gas are likely to reach temperatures well in excess of the tempering temperature.  Distortion of the races is, therefore, one potential cause of failure and, with unpredictable temperature gradients as a separate factor, it is essential to fit bearings with above-normal internal clearance. Lubricants are available which have been proved to operate for a number of hours with bearing temperatures between 400ºC and 450ºC, but they are not suitable for re-lubrication by normal means, and are extremely expensive.  More conventional lubricants with high drop point are also available and will, therefore, be preferred for emergency temperatures up to 250ºC or 300ºC.

The aluminium alloys used for many motor enclosures are not the most stable at elevated temperatures, so if failure is not to occur due to the rotor rubbing against the stator, aluminium enclosures have to give way to ferrous at some point in the range. Even then precautions such as dowelling of stator cores have been shown to be necessary.

The options for the organic insulating materials in a motor are numerous, but a discussion of these would be out of place in this paper.

Two points should, however, be made: firstly that specifiers often call for Class H or even Class C insulation, not realising that the ability of materials to meet these criteria has little relevance to the manner in which they degrade at the much higher temperatures needed for smokespill duties. Specifications should define the functional requirements and only very rarely how it is thought that these should be met. Secondly, it is possible to use widely-available and low-cost organic materials for emergency conditions of up to 1 hour at 300°C; above this, however, it is essential to use exotic materials that are very much more costly, and not so readily available. Preventive maintenance would pose some difficulties, as an associated disadvantage to the client who believes he has to allow for temperatures in excess of 300°C.

### 4.4 Selection of fan type : bifurcated or Motor-in-airstream

Consideration of materials that are currently available and their cost, in relation to the temperature/time specifications listed, lead to a simple selection guide as follows:

| Category Codes | Fan Type |
|---|---|
| 650/1 | Motors are not considered to have sufficient reliability for operation in airstream : bifurcated fan designs should be adopted. |
| 400/2 | Either bifurcated or motor-in-airstream designs may be considered : the latter will require costly materials and impose greater demands for maintenance but will need to be adopted if the fan cannot be mounted in an area well away from the fire zone. |
| 300/1 300/0.5 | Motor-in-airstream designs will usually be the preferred choice : although non-standard materials will usually be needed, the effect on cost and maintainability should not be severe. |
| 150/5 | Motor-in-airstream will be the natural choice, since conventional materials will suffice. (This does not mean that every standard motor will be satisfactory - the use of thermoplastic cooling impellers and terminal blocks, for example, needs to be checked). |

### 5 CERTIFICATION REQUIREMENTS

Views on the need for product certification vary widely, and the UK has not reached the same position as have West Germany or France where certification, by TuV and CTICM respectively, is in most situations a pre-condition to selling fans for smokespill duties. Nevertheless, there is now a proposed UK standard concerned with the testing of such fans, prepared by the Smoke Ventilation Association of HEVAC, and this is now with BSI for consideration as a National Standard.

Product certification, whether it be by an independent authority or on a self-certification basis, will need to consider the relationship between a single test result and the probability of survival of other similar products. There is considerable statistical uncertainty concerning the spread of the lives of a sample of supposedly identical machines - this is increased when certification embraces a range of sizes of fan, and is compounded by the uncertainty arising when the product may have been in service for a long time prior to an emergency occurring.

The data available to the authors has led us to require that if a single test result is to be used, then the measured life should be at least twice that which is specified, and only if a sample of three or more identical machines has been tested should a lower margin of safety be considered. Table 6 indicates the spread that has been measured in our tests. Although this is a self-penalising attitude, we believe it to be better engineering than placing reliance on a single test result. Indeed CTICM have recognised the aspect of uncertainty sufficiently to require that their tests are extended 20% longer than the required life, i.e. a 2 hour claim requires a tested life span of 2hrs 24mins.

### 6 INSTALLATION AND MAINTENANCE

It is only necessary here to note that to allow for a compatible installation, the fan manufacturer must provide terminal arrangements that can accept mineral-insulated copper-covered cables - though other types of cable, with polymeric insulation suitable for this emergency duty, can now be obtained. Obviously any ancillaries such as flexible connectors and anti-vibration mounts need to have the same thermal capabilities, in addition to their normal functions.

Maintenance is a more difficult matter as the use of a fan for a normal duty means that if an emergency does occur after several years of operation, the probability that it will survive the specified conditions is inevitably reduced. To counter this uncertainty requires, at best, an unsatisfactory degree of extrapolation of known temperature/life relationships. One recommendation currently included in the Maintenance Instructions notes "The Engineer responsible should recognise the advisability of a major re-fit after operation for 30 000 hours. This would involve rewinding with materials of equivalent thermal endurance to those in the motor supplied originally, and replacement of bearings. Further guidance on the time span prior to re-fit can be given on receipt of full operational information, duty cycle, and nameplate data."

This is an area where further work is undoubtedly called for, but meanwhile it may be that a consensus of engineering judgement can be arrived at, along lines such as these.

Table 6   Variability of motor life at specification temperature

| Test Group | No. in Sample | Test Temp. | Motor Speed | Life at Test Temp. (hrs) Min. | Max. | Mean | Std.Dev'n |
|---|---|---|---|---|---|---|---|
| A | 3 | 400 | 2900 | 5.7 | 7.9 | 6.5 | 1.0 |
| B | 3 | 400 | 1450 | 3.0 | 3.8 | 3.4 | 0.34 |
| C | 2 | 300 | 2900 | 3.2 | 7.1 | 5.1 | 2.0 |

## 7   CONCLUSIONS

In the design of buildings, it is now widely recognised that it must be made patently clear that emergency situations have been allowed for, even if they are unlikely ever to occur.  The application of fan systems to this situation has been shown to have a sound basis in the work carried out by the Fire Research Station and others.

   In the selection of fans for such systems, it has also been made clear that designs can be provided to meet all duties that have so far been demanded.  Specifiers need to recognise, though, that it would be bad engineering to call for temperature/time combinations that are more onerous than can reasonably be postulated for a given duty - certainly costs can increase sharply if this is done.  Of equal importance is the recognition that maintenance arrangements need engineering consideration and a high degree of integrity in the way they are carried out in service.

REFERENCES

(1)  THOMAS, Hinkley, Theobald, Simms.  Fire Research Technical Paper No.7 1964

(2)  MORGAN, H. P.  B.R.E. Information Paper 19/85

(3)  MORGAN, H. P.  B.R.E. Paper No.34 1979

(4)  BUTCHER, E. C.  Fire Progression Spread and Growth/Fire Size. October 1987

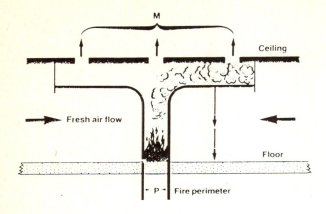

Fig 1    Production of smoke -- single affected space

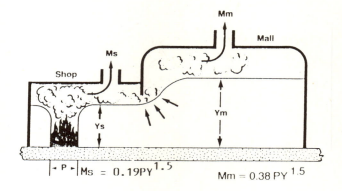

$$M_s = 0.19PY^{1.5}$$

$$M_m = 0.38PY^{1.5}$$

Fig 2    Production of smoke in shopping malls

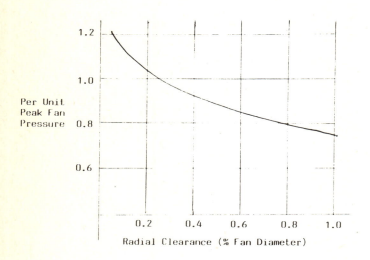

Fig 3    Effect of tip clearance in axial fan on pressure

Fig 4    2.8 m fan after HT250/1 test

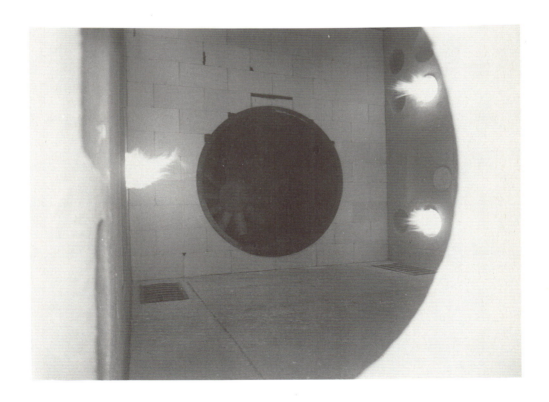

Fig 5    1.25 m bifurcated fan on test at 620°C [courtesy of TüV, Munich]

# C401/022

# Installation 'know-how' as regards fan installation applications for rapid transit systems

Eur Ing **I J COCKRAM**, CEng, FIMechE, FCIBSE, FInstE, MASHRAE
Locke Cockram and Associates Limited, London
**T MAHONEY**, Assoc CIBSE
London Underground Limited, London

SYNOPSIS A non-technical paper which outlines the ground rules for a basic common sense approach to the ongoing problems of environmental and smoke control for both existing and new Transit Systems.

## 1 INTRODUCTION

Installation 'KNOW-HOW' can really be interpreted as another word for EXPERIENCE and obviously only achieved through long term 'trial and error' – especially in the particular field of mechanical ventilation for Rapid Transit Systems. However, as far as Smoke Control is concerned, this is a separate issue altogether and is therefore mentioned later on.

For almost a century, London Underground has been using this 'know-how' and in the process has integrated most types of mechanical ventilation plant into it's system, thereby endeavouring to control the environment for it's passengers (customers) and staff alike – as well as endeavouring to be one of the foremost advisers to the world's major transit authorities.

There have been many papers presented on fan design and technology, with every manufacturer, like the car industry, trying to improve over his competitors, even with minor improvements, so as to make his fan different from the rest.

A fan could be defined as a mechanical device for the movement of air against a specified resistance to airflow: This statement is admittedly fundamental, but nevertheless with the aid of ever advancing technology (including computers), it is all too easy for manufacturers and designers to partially forget the primary objective.

This paper tries to set out and expand on previous papers, also at the same time giving the latest views on improvements that are being undertaken by certain ventilation engineers, so as to improve safety standards... having regard to ventilation, control of smoke, passenger evacuation and entry to the system for emergency personnel etc..

## 2 THE OPERATION OF A FAN AND THE ATTENTION REQUIRED TO MINIMISE IT S EFFECT ON THE ENVIRONMENT

The 'Tunnel Cooling' fans (as they are called), serving the 'Tube System' range in duty from 10 to $60 m^3/s$ and serve two purposes: Firstly to control the environment of the tunnel in removing heat at night when the trains are not operating and secondly, to provide reasonable air movement for passengers when a train is stopped in a section of tunnel.

Over the past four years, the duty of these fans has out of necessity, had to be investigated with the object of increasing their output by 300 per cent. Since these fans are the only positive means for smoke removal, the operating controls and who operates same, has had to be given a greater importance.

London Underground's Engineers have for many years prided themselves on being able to disguise the large tunnel fan plant rooms throughout the City and at the same time operate such machines without noise pollution or aesthetic degradation. Any engineer will know that if the power of a machine is increased by 300 per cent, using the same housing and air route, then immediately one has a problem with regard to noise emission from the point of discharge .... especially when located in a large city where buildings are 'everywhere'. The engineers, even with their installation 'know-how' have to consider the noise problem very carefully for any transit system – especially when endeavouring to update it's mechanical ventilation plant in order to meet the future criteria requirements.

Any large fan must have in this day and age the capability of electrical reversal. If one takes a 'normal' high efficiency axial flow fan and operates it in the reverse mode, then the efficiency will obviously be reduced e.g. 85 per cent normal running, 45 per cent in reverse operation – this being mainly due to the aerofoil shape of the fan blade: It should be mentioned at this point however, that these percentages will in fact vary greatly from manufacturer to manufacturer ... sometimes the difference can be as low as 15 per cent.

If 50 per cent of the blades are reversed, then the duty/airflow of the fan could possibly be the same in either direction, BUT the overall fan efficiency will drop to say 50-60 per cent in BOTH directions, with subsequent power loss and an increase in noise due to blade turbulence.

In the early days of fan technology, one usually put the biggest size of fan in the biggest hole size available! - fortunately the result was usually effective. The fan systems designated for the future obviously have the advantage of computerised technology, so as to give a better forecast on the result (but not in all cases). Nevertheless it is still far better to put a large low-revving fan into the system, than a high speed 'screamer' so as to overcome the noise and turbulence.

On the assumption that the mechanical plant has been designed for greater output, with minimal noise increase, then the 'air noise' caused by the increased velocity through the ducts and tunnels breaking-out at surface level, needs to be addressed. Attentuation in an 'old' Transit system running in a heavily populated commercial/domestic area, is unfortunately fraught with problems.

The more air that is moved, the more dust that is carried with it and within a very short period of time, the attenuator loses it's effectiveness and due to the nature of the dust, a fire hazard could develop. Consequently regular cleaning is of the utmost importance.

It takes approximately 6 months to re-design, remove and install a new fan system. A fan could be running 24 hours/day alongside a domestic premise having a constant noise level rise of 3 dBA above the lowest background level for many years and without problems. For this 6 months transition period, the fan system is out of service - All is quiet! The new fan is then started-up and with only 2 dBA above the background noise level all sorts of commotion develops with Solicitors and Public Health Officers acting on behalf of Complainants ... the human element is therefore always worthy of consideration!

The next problem, assuming the fan is as efficient as it can be in meeting the duty, is quietness in operation, so as not to arouse further public complaints: The point of discharge must be assessed. As the objective of the fan is mainly to move warm air from the tunnels, which is normally laden with dust as mentioned earlier and although non-toxic or irritant, it does contain a high element of oily/carbonaceous particulates, which if discharged against an adjacent property or window could, after a very short period, cause staining of the premises with all the inherent complaints.

However, the biggest problem to be thought out is that knowing the fan could be used for smoke removal, then it is absolutely essential that the discharge point is away from openable windows - so that any smoke being discharged into the atmosphere from down below, does not enter another building - with the obvious consequences.

## 3 FAN DESIGNS SO AS TO OVERCOME TRAIN ASSISTED AIR MOVEMENT

In the modern-day transit system, trains are required to travel at the maximum allowable velocity (including both acceleration and deceleration) through the tunnels. With good suspension and track 'canting', air space around the train can unfortunately be kept to a minimum. The air volume caused by the 'Piston' action of the trains' movement can activate 4 to 5 times the amount of air normally required to control the environment by fan assistance - but obviously not in such a positive and controlled manner, since the movement of trains can be said to 'pulse' the air backwards and forward as compared to fan assistance.

The over-pressure/volume so caused by the piston effect, obviously has a detrimental affect on the fan and motor... in that it can either attempt to speed-up and assist the rotation of the fan, or 'stall' it in reverse, with the subsequent over-loading of the motor. The pressure surge however, usually lasts for only a few seconds and in the past, motors fitted to Axial fans were over-rated by up to 40 per cent so as to cope with this situation.

When centrifugal fans are installed, they have backward curved blades, thereby producing a non-overloading characteristic.

Modern fans, because of the requirement for full reversability regarding smoke control, have to be designed, so that they are equally reliable in adverse flow conditions, but at the same time, economical with respect to electrical consumption during every day normal operation.

A recent design 'innovation' being the anti-stall ring and which has proven to be totally effective for the aforementioned problem.

Another more basic solution is the fluid drive coupling, which is located between the rotor and motor, with 'slipping' taking place when the air flow is reversed by train movements.

Non-return dampers used to be fitted to the fan inlet so as to prevent train action reversal occurring, but with the advent of bi-directional fans to remove and/or control smoke such dampers are no longer allowed.

## 4 USE OF COMPUTER AIDED DESIGN MODELLING FOR THE CALCULATION OF AIRFLOWS PRODUCED BY COMPLEX TRAIN MOVEMENTS

It is quite a lengthy calculation to determine the airflow around the labyrinths within a station for one given circumstance, in that the air being moved by the piston action of a train is continually accelerating and de-celerating within it's own tunnel and similarly in the 'other direction' tunnel (which has it's own train movements).... see Fig. 1.

Consider then the worst case and the complexity of 3 lines, with 6 trains all entering and departing a common station at various intervals. How much air one train moves can be calculated by obtaining the following criteria:-

- i) Time of train journey from adjacent station.
- ii) $Accl^n/Decel^n$ also maximum speed.
- iii) Tunnel dimensions.
- iv) Blockage ratio $\dfrac{\text{Nett X-Sect Area Train}}{\text{Nett X-Sect Area Tunnel}}$
- v) Distance between adjacent stations.
- vi) Locations of adits/cross-passages.
- vii) Aerodynamic drag.
- viii) Length of train.

In conjunction with the aforementioned information, a speed/time graph or graphs can be drawn for the services.

If a train travels a certain distance, which is then subsequently divided by the 'lapsed' time - the result of which is multiplied by both the nett tunnel cross-sectional area and the blockage ratio, then a flow rate can very simplistically be derived. This is for one train arriving or departing the station and ignoring the cross-passages (adits) between the two running tunnels: Consider then a correct calculation for 6 trains and the variables that can exist with different speeds, tunnel sizes, distances and train performance characteristics. For the worst condition it should be assumed that all trains either arrive or depart the station at the same time. Such a 'scenario' will obviously produce the maximum envisaged airflow within the station under assessment.

To now calculate the air speeds within the ticket halls, platform areas, passenger subways and the like ... knowing from which tunnel portal/headwall the air moves and of what volume, now raises the simple mathematical calculation to that requiring the assistance of a computer. It can and has been undertaken by laborious manual and pro-rata calculations based on the assumption that friction losses around the station are equal, so that air velocities are similarly obtained as for a pipe or ductwork calculation.

Imagine now having worked out by careful means all the airflows and velocities around the station - then another variable i.e. speed change or even a different type of rolling stock is introduced. One can now see the advantage of computer application and it's use in saving time etc..

It must be appreciated that although computers are an aid to technical know-how ... they can be wrong and it is up to the design engineer to use his experience in assessing the answers and cross-checking accordingly.

## 5 TELEMETRIC CONTROL FROM A CENTRAL LOCATION

Over the past two years, London Underground's engineers have undertaken a programme of introducing into the system a telemetric method of fan control, using the signals through standard British Telecom lines to both monitor and control the operation of the fans. This requirement has been emphasised with the importance now being placed on the fans as a means of removing smoke (within the best of their capability). Most tunnel cooling fans as they are called, are installed mid-way between stations and have a hard-wired remote control link, which is taken back to a location in one of the flanking stations.

For the system of control to function correctly, it relies on the maintenance engineer to remember to switch the fan control BACK to the 'Remote' position. Failure to do so, would mean that in the event of a fire/smoke related incident, the station staff or the Fire Brigade would try to operate the fans remotely, or think they had, but the fan would still operate in the mode set under local control! Consequently the station staff (or others) would have to walk some 1.5 km from the station to operate the fan correctly - by which time it would surely be too late.

The telemetric control system still allows the engineer to take control of the fan, but as soon as he does, an alarm is registered as a 'MAN-ON-SITE' on the monitor in the Ventilation Engineer's control centre. Whilst in the Man-On-Site mode, the fan is still not capable of being operated remotely (for obvious safety reasons): The alarm will stall register in the control centre.

Sensors in and around the fan, send back additional information on the state of the fan viz:-

    Running or Stopped
    Tunnel Air Temperature
    Fan Amperage
    B T Line Status
    Mains Supply ... Healthy/Failed
    Battery Back-Up Failure
    Man-On-Site

With this information, economical operation of the fan can be gained by running the fan within a designated temperature band, i.e. above 25 Deg.C - fan is switched ON and below 15 Deg.C the fan is switched OFF. Each of the sensors generates an alarm back to the control centre, should there be a change-of-state outside the designated limits.

Safety for the Fan Maintenance Engineer is still paramount. Even when the fan is running in it's normal mode, large notices on the fan and control panel, warn the Operator ...THIS FAN IS LIABLE TO BE OPERATED FROM A REMOTE LOCATION - TURN

KEY SWITCH TO 'MAN-ON-SITE' AND ISOLATE MAINS BEFORE CARRYING OUT MAINTENANCE.

Such a notice also warns the casual visitor to the site, who could pass a fan and have it suddenly start-up before him.

## 6 SMOKE LOGGING IN TUBE TUNNELS

The running tunnels for any rapid transit system are obviously never level, since they have to be tunnelled over and around foundations, sewers and other such protuberances etc.. A stationary train on fire in a tunnel section presents the worst hazard, in that smoke from the fire can be generated and pushed by the fire load into the adjacent tunnels ... where it cools, slows down and then builds up in density to smoke log section of tunnel or indeed a train.

Fans can be used with great effect in removing smoke from the tunnels and their capacity to carefully move air, can be worked out to ensure that when in operation, a minimum airflow of 2.4 m/c is maintained along the nett tunnel cross-section (not the Annulus). This velocity criteria has been proved capable of clearing smoke for a number of practical tests.

The volume of these fans are usually in the order of 200 $m^3$/s and preferably located in such a manner that a 'Push-Pull' method can be utilised and a positive air displacement is achieved and additionally, a degree of stand-by facility.

It should be mentioned at this stage, that it is highly desirable, if not essential, only ONE train is allowed into a Station to Vent Shaft, or a Station to Station Section, since, due to the large cross-sectional area of the train (occupying some 60 - 62 per cent of the tunnel), then there is every likelihood of smoke from one train entering another 'Upstream', when it passes between the train sides and the tunnel wall ... Due to an 'Over-pressure' development[1] ... see Fig. 2.

## 7 THE APPLICATION OF FANS FOR THE PROVISION OF A SAFE AND SMOKE-FREE MEANS OF ESCAPE FOR PASSENGERS AND/OR ENTRY ROUTE FOR EMERGENCY PERSONNEL

Where peoples' lives are concerned, it is important to stress and separate out such requirements.

Safe - means that in the event of a fire/smoke related incident in a station, then the dedicated route used for evacuation should be safe for use, with no possibility of cross-contamination of smoke from other lines.

Smoke Free - this means that routes used for evacuation shall have a mechanical ventilation plant for use when dealing with:-

    i) Station Fire - To provide a positive movement of fresh air throughout the whole length of the escape route ... from platform to surface and which will inhibit the flow of smoke, but NOT increase the burning rate by 'feeding' the fire with oxygen.
    ii) Tunnel Fire - To provide sufficient air-flow to carry smoke being released from a fire on a train, to the opposite direction of evactating/de-training passengers.

A rather contentious specification requirement is that of the temperature and duration to which

...n is to operate during emergency conditions :
...me specifications call for:-

149 Deg.C for 1 hour.
149 Deg.C for 2 hours.
250 Deg.C for 1 hour.
250 Deg.C for 2 hours.

Others have been seen to call for 600 Deg.C
with a duration of 4 hours! (obviously a special
motor, since this is usually the 'problem part').
All that is sufficed to quote is surely a
practical need ... say 149 Deg.C for 1 hour and
on some special projects, a maximum of 250 Deg.C
for 1 hour.

REFERENCES

(1) COCKRAM. I. J. Smoke removal in tunnels and
    it's effect on the safety environment of
    rail cars. BHRA 6th International Symposium
    on the Aerodynamics and Ventilation of Vehicle
    Tunnels. Durham. September 1988.

ACKNOWLEDGEMENT

The Authors wish to thank the Directors of
London Underground Limited for their permission
to publish this paper.

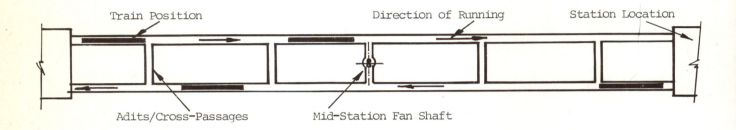

Fig 1

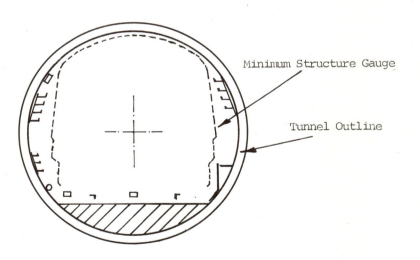

CROSS SECTIONAL AREA
THROUGH RUNNING TUNNEL

Fig 2

# A comparison of the performance and sound levels of jet fans measured in a laboratory and in two road tunnels

C L B NOON, DFH, CEng, MIEE, MCIBSE and T W SMITH, BSc, CEng, MIMechE
Woods of Colchester Limited, Colchester

SYNOPSIS This paper describes the test methods used to measure air volume, thrust and open inlet and discharge sound power levels under laboratory conditions and discusses the justification of their use.

When installed in a tunnel, the jet type fan does not appear to perform as effectively as laboratory tests would suggest. It has been established that the siting of a fan close to a wall increases the losses.

The methods used for performing site tests are described and the difficulties encountered. The paper studies the results of airflow and noise measurements taken in two road tunnels and compares these with a theoretical assessment to establish an installation factor.

## NOTATION

| | | |
|---|---|---|
| F | fan thrust | N |
| $P_f$ | pressure rise by jet fan | Pa |
| $\rho$ | air density | kg/m³ |
| $A_f$ | cross sectional area of fan | m² |
| $A_t$ | cross sectional area of tunnel | m² |
| L | length of tunnel | m |
| $V_f$ | fan outlet velocity | m/s |
| $V_t$ | average tunnel velocity | m/s |
| H | distance between fan axis and tunnel ceiling | mm |
| D | jet fan diameter | mm |
| $P_s$ | system resistance | Pa |
| $D_h$ | tunnel hydraulic diameter | m |
| $\dot{m}$ | mass flow | kg/s |
| a | area | m² |
| v | velocity | m/s |
| Q | volume flow | m³/s |
| V | volume of tunnel | m³ |
| T | reverberation time | secs |
| $l_e$ | Entry loss factor | |
| $l_o$ | exit loss factor | |

## 1  INTRODUCTION

In recent years there has been a marked trend towards the design of tunnel ventilation systems which provide a low first and operating cost without sacrifice of reliability or system performance.

This specification can be met using the longitudinal ventilation system whereby the tunnel itself supplies all the fresh air requirements for pollution control without the need for additional air ducts. The fresh air is induced at one end and discharged at the other end by the action of high speed axial flow fans. These fans produce a high velocity jet which entrains a much larger flow through the tunnel cross sectional area. Where traffic is free moving above 20km/hr, the piston action due to vehicular movement, is adequate for pollution control on most tunnels under 500m in length.

The fan or combination of fans has to develop a total pressure rise which will balance the system pressure requirements as established by conventional methods for the specified flowrate.

The fans are normally hung from the roof of the tunnel in pairs as shown in figure 1a. Where headroom is limited, such as in a 'Cut and Cover' type tunnel, they maybe fixed on either side as shown in figure 1b.

The efficient transfer of the momentum energy of the fans to the tunnel airstream is influenced by the proximity of the fan to the adjacent wall or roof or both. Also this efficiency is effected by the closeness of fans if located side by side and to the distance between successive fan locations down a tunnel. Some results of work from laboratory models have been published on the influence of walls adjacent to a fan. (1,2) Rule of thumb data has indicated a minimum spacing between fan locations down a tunnel.

## 2  PERFORMANCE MEASUREMENT

To establish the performance of a jet fan, the following information is required :

(a)  Volume flow
(b)  Axial thrust
(c)  Input power
(d)  Sound level

### 2.1  Volume flow

The most obvious theoretical choice of test method for measuring the volume flow of a jet fan would be to use the BS848:Part 1:1980 Type-A chamber test method with the fan mounted half way into the chamber. However, the practical problems associated with testing in this manner are that it is necessary to remove the fan from the thrust rig and install it on the chamber between tests, so that the pitch angle for the thrust test and volume measurement are exactly the same. The preferred method is to use a conical inlet in accordance with BS848. This type of inlet has an inlet loss coefficient which is almost identical to that of the bellmouth inlets used on jet fans, although, of course, it is necessary for the designer to confirm that this is the case. It therefore has the same loss and will result in the volume flow being the same during the thrust and volume flow tests. The conical inlet is mounted in place of the bellmouth inlet and connected to the silencer, as shown in figure 2. The volume flow is calculated using the static pressure change from ambient to side pressure tappings. The clearance around the inlet must be in accordance with BS848 to ensure that the velocity distribution around the inlet is not affected by obstructions.

## 2.2 Thrust

The theoretical thrust of a jet fan is :

$$= \Sigma \, \dot{m} \, \Delta V$$
$$= \Sigma \, \rho \, q \, \Delta V$$
$$= \Sigma \, \rho \, a \, \Delta V^2$$
$$= \int \rho \, 2 \, \pi \, r \, dr \, \Delta V^2$$

If the velocity is constant :

$$= \rho \, \pi \, R^2 \, \Delta V^2$$
$$= \rho \, Q \, \Delta V$$

Which is the conventional means of calculation. However, velocity clearly does vary with radius although the relationship is complex and cannot be established theoretically. It is, therefore, necessary to measure the fan thrust.

Fortunately action and reaction are equal and opposite so the thrust imparted to the air can be established by measuring the reaction on the fan. The method of doing this is shown in figure 2. The fan is mounted on a frame supported by four linear bearings. These bearings have a static friction coefficient of .0025, which is, in practice, probably reduced by the unsteady flow conditions causing a slight fluctuation in thrust and hence movement. For a typical 1 metre jet fan, with a weight of 850Kg, and a thrust of 1000 N the friction force would be 21 N. As the fan thrust fluctuates, this friction force does not affect the mean measurement but provides a degree of damping.

Linear bearings work very well in practice as they make the test method stable by restraining non axial forces without adding any significant axial friction forces. They overcome the problems of suspended rigs of ensuring that the suspension is absolutely vertical. They also provide a means of ensuring that the rig is horizontal, which can be checked by measuring the force needed to move the fan in each direction.

The thrust force is measured by means of a load cell which is used to restrain the axial movement of the fan.

Care is needed over siting of the fan and the size of building where the rig is installed. Jet fans entrain the movement of a large amount of air. If the air is already moving as it approaches the fan, then the thrust, which is the

change of momentum of the air, will be reduced. The discharge velocity from the jet fan will typically be 20-35 m/s², a velocity of 1 m/s upstream of the fan will reduce the thrust by 3-5 per cent. For fans tested in a restricted space, it is often necessary for the fan to discharge to atmosphere or to point into a corner to reduce the induced velocities. Equally it is necessary to test under steady state conditions; it is no use switching the fan on and taking the instantaneous maximum reading before normal flow conditions have been established.

It is interesting to compare the measured thrusts with theoretical estimate based on a uniform velocity throughout the full duct area. The measured values vary between 85 per cent and 105 per cent of theoretical, depending on the velocity distribution and the internal fan and silencer configuration.

## 2.3 Sound levels

The measurement of sound levels is, also, quite straight forward. The method used complies with BS 848:Part 2, see figure 3. The thrust and air tests are performed in a large laboratory which is fairly reverberant. The laboratory is calibrated by using a sound source of known power and the reverberation levels checked to show there is a consistent sound pressure level in the measurement region. Measurements are made at an angle of 45º to the fan axis at the inlet and outlet, by turning the fan through 180º. It is necessary to measure both inlet and outlet sound levels as they differ, with the outlet levels being about 2dB higher and with a slightly different spectrum. To calculate the sound power level the difference between the measured sound pressure levels for the known source and the fan being tested are added to the sound power level of the known source in each octave band. The inlet and outlet levels are added to give the total sound level emitted by the fan. For reversible fans, the measurements are made in both direction as even though the volume and thrust figures for a truly reversible fan may be virtually identical in both directions the sound level would normally be different because there are struts supporting the motor which will be upstream of the impeller in one direction but downstream in the other.

## 3 SYSTEM CALCULATIONS

## 3.1 Fan pressure rise

The pressure rise developed by a Jet fan when mounted in a tunnel is :

$$P_f = \rho \, V_f^2 \, A_f \left( 1 - \frac{V_f}{V_t} \right) K_1 \, K_2 \ \text{Pa}$$

$K_1$ = Fan thrust factor varying with fan outlet velocity distribution and fan configuration. Laboratory tests indicate values between 0.85 and 1.05 - dependent upon blade and fan design.

$K_2$ = A thrust loss factor dependent upon :

(a) The distance between fan axis and wall and/or ceiling.

(b) The distance apart of adjacent fans.

(c) The distance between successive fans down the length of a tunnel.

(d) How close the first fan is mounted to the tunnel entry portal.

(e) How close the last fan is mounted to the exit portal.

## 3.2  System resistance

The total pressure required to maintain the required airflow through the tunnel, is the summation of the following :

(a) The tunnel entry loss factor. Normally 0.5 but can be higher where the portal protrudes outside the hillside.

(b) The loss due to the tunnel wall friction. Normally taken as 0.024 which allows for some appendages such as road signs.

(c) Outlet tunnel loss factor - normally 1.0.

(d) The meteorological effects, which can be the barometric difference between portals and also the pressure difference due to the wind. The former is normally small but a pressure of 20Pa is generally allowed for wind if more specific data is not available.

Hence the system resistance is given by the equation :

$$P_s = (l_e + l_o + f \frac{L}{D_h}) \frac{1}{2} \rho V_t^2$$

## 3.3  Fan Selection

The number of fans required is :

$$N = \frac{P_s}{P_f}$$

The value of $P_f$ is determined by the size of fan and its blade angle. The fan size chosen depends on a number of factors, namely :

(a) Space available for the fan.

(b) Overall efficiency in terms of newtons thrust per fan input kilowatts.

(c) Distribution over the tunnel area.

(d) Length of tunnel.

(e) Steps of fan control.

(f) The lowest cost of installed fans including cabling.

(g) Noise.

## 4    SITE JET FAN PERFORMANCE

It is seen from the above that the resultant performance of the jet fans when operating together in a tunnel, can be considerably lower than that achieved by laboratory measurements.

The calculation of the air quantity required to limit the pollutants to an acceptable level, are established for a specified vehicular flow through a tunnel. Alternatively the maximum flow rate maybe determined by a minimum tunnel velocity to control the smoke layering as the result of a fire. Under these fire conditions, an assumption would be made as to the number of stationary vehicles and the resultant drag associated with the air velocity specified. The percentage and their distribution in the tunnel of trucks can have a great influence on the drag.

The commissioning of the jet fan system of ventilation has its difficulties and the only practical flow test that can be carried out is to measure the airflow through an empty tunnel. This flow rate can then be compared to a theoretical calculation of the pressure loss through the tunnel and the total pressure rise of the jet fan installed.

The site tests result of airflow have been evaluated for the following UK tunnels :

Penamaenbach

Tunnel details :  2 lanes
        Length          658m
        Area            66.7m²
        Hydraulic Dia   8.57m

Jet fan installed:12
        Size            630mm fully
                        reversible
        Speed           2950 rev/min
        Motor rating 11kW

Laboratory test result:

|  |  |  |
|---|---|---|
| Air volume | Form A | 9.0 m³/s |
|  | Form B | 9.1 m³/s |
| Air velocity | Form A | 30.7 m/s |
|  | Form B | 31.0 m/s |
| Thrust | Form A | 307 N |
|  | Form B | 326 N |

Measured / theoretical thrust ratio:

        Form A    0.93
        Form B    0.96

Sound power levels:

One diameter length of silencer either side of fan.

|  | Inlet | Outlet |
|---|---|---|
| Form A | 103.2dB | 101.0dB |
| Form B | 97.3dB | 99.3dB |

Re: $10^{-12}$ Watts

Site airflow measurements :

The velocity readings were taken from a digital micromanometer connected to a pitot tube.

Results of 29 readings taken over the area - evenly distributed for no fans operating and all 12 fans operating.

        Fans on, average velocity    3.67 m/s
        Fans off, average velocity   0.38 m/s

The corrected value of tunnel velocity for the effect of the wind is :

$$V_t = \sqrt{V_m^2 + V_w^2}$$

$V_m$ = measured velocity with fans operating
$V_w$ = measured velocity with no fans operating

$$= \sqrt{3.67^2 + 0.38^2}$$
$$= 3.69 \text{ m/s}$$

System resistance :

| | | |
|---|---|---|
| Velocity pressure | $= \frac{1}{2} \rho V_t^2$ | |
| | $= 8.08$ Pa | |

| | |
|---|---|
| Inlet loss (K=0.5) | $= 4.04$ Pa |
| Outlet loss (K=1.0) | $= 8.08$ Pa |

Friction loss (f=0.024) $= f \dfrac{L}{D_L} \ \frac{1}{2} \rho V_t^2$

$$= \frac{0.024 \times 658 \times 8.08}{857}$$
$$= 14.9 \text{ Pa}$$

| | |
|---|---|
| Total system resistance | $= 4.08 + 18.7 + 14.9$ |
| | $= 27.15$ Pa |
| Effect of wind | $= 0.29$ Pa |

At the time of the site measurements, two workmen's huts were located in the tunnel. Each had a face area of approximately 2.5 m². This represented a loss value of an estimated 1.5 Pa hence the estimated system resistance.

$$= 27.15 + .29 + 1.5$$
$$P_s = 28.9 \text{ Pa}$$

Fan pressure rise :

$$P_f = F_m \ \frac{1}{A_t} \left( \frac{V_f - V_t}{V_f} \right)$$

Where F = measured fan thrust – Newtons

$$P_f = \frac{307}{66.7} \left( \frac{30.7 - 3.67}{30.7} \right)$$

$$= 4.05 \text{ Pa}$$

The apparent installation factor

$$K_2 = \frac{P_s}{n \ P_f}$$

n = number of fans

$$K_2 = \frac{28.9}{12 \times 4.05}$$

$$= 0.59$$

Saltash Tunnel

| Tunnel details | : | 3 lanes |
|---|---|---|
| | | Length   411m |
| | | Area      92m² |
| | | Hydraulic Dia 9.95m |

| Jet fans installed - 6 | : | Size 1000mm fully reversible |
|---|---|---|
| | | Speed  1470  rev/min |
| | | Motor rating 22kW |

Laboratory test results :

| | | | |
|---|---|---|---|
| | Air volume | Form A | 24.7 m³/s |
| | | Form B | 24.7 m³/s |

| | | |
|---|---|---|
| Air velocity | Form A | 31.45 m/s |
| | Form B | 31.45 m/s |
| Thrust | Form A | 836 N |
| | Form B | 833 N |

Measured / theoretical thrust ratio:

| | |
|---|---|
| Form A | 0.9 |
| Form B | 0.89 |

Sound power levels:

1m length silencers either side of fan

| | Inlet | Outlet |
|---|---|---|
| Form A | 91dB | 96dB |
| Form B | 90dB | 90dB |

Re: $10^{-12}$ Watts

Site airflow measurements :

Fans blowing west to east.  32 readings of velocity measured over the tunnel area with a vane anemometer at two sections.

Average velocity = 5.2 m/s

System resistance :

| | |
|---|---|
| Inlet and Outlet loss | $= 1.5 \times 16.2$ |
| | $= 24.3$ Pa |
| Friction loss | $= 16.1$ Pa |
| Total system resistance | $= 40.4$ Pa |
| Effect of wind | $=$ negligible |

Fan pressure rise :

Fan running Form A

$$P = \frac{836}{92} \frac{(31.45 - 5.2)}{31.45}$$

$$= 7.58 \text{ Pa}$$

Number of fans operating = 6

$$K_2 = \frac{40.4}{6 \times 7.58}$$

$$K_2 = 0.88$$

Fans blowing east to west

32 readings of velocity measured over the tunnel area with a vane anemometer at two sections.

Average velocity  =  4.2 m/s

System resistance :

| | |
|---|---|
| Inlet and outlet loss | $= 1.5 \times 10.6$ |
| | $= 15.8$ Pa |
| Friction loss | $= 10.5$ Pa |
| Total system resistance | $= 26.3$ Pa |

Fan pressure rise:  Fan running Form B

$$P = \frac{833}{92} \left( \frac{31.45 - 4.2}{31.45} \right)$$

$$= 7.85 \text{ Pa}$$

Number of fans operating = 6

$$K_2 = \frac{26.3}{6 \times 7.85} = 0.56$$

## 5 SOUND LEVELS

Extensive sound level measurements were taken in the two tunnels with all fans operating. The reverberation times were also recorded. It is possible to compare the predicted levels in the tunnels with those measured.

### Penamaenbach Tunnel

Sound level measurements were taken at 30 metre intervals, at a height of 1.5 metres above the road surface, along the tunnel centre line, all the fans running in the Form A direction.

The average of the measured levels in each octave band is as follows:

|  | 63 | 125 | 250 | 500 | 1K | 2K | 4K | Hz |
|---|---|---|---|---|---|---|---|---|
| Sound pressure level | 85 | 88 | 84 | 86 | 86 | 83 | 75 | dB |

The background was 13 dB lower as a minimum.

Estimated sound level in tunnel :

This is evaluated by calculating the reverberant level in the tunnel. A calculation based on the direct level is not considered since the site measurement of the direct level is only 2dB higher than the minimum level measured in the tunnel at any point down the tunnel.

The resultant sound pressure is :

$$= \text{sound power level (SWL)} + 10 \log \left( \frac{4T}{.161 \, V} \right)$$

The total sound power level from all fans operating is :

$$= 10 \log R + \text{SWL (1 fan)}$$

where R is number of fans operating, hence for 12 fans = SWL + 11

Laboratory test data:

The sound level measurements recorded in the Laboratory on one of the fans fitted on site.

|  | 125 | 250 | 500 | 1K | 2K | 4K |  |
|---|---|---|---|---|---|---|---|
| Inlet | 98 | 97 | 95 | 93 | 95 | 88 | dB |
| Outlet | 94 | 96 | 94 | 94 | 93 | 86 | dB |

These are sound power levels ref: $10^{-12}$ Watts.

When installed in the tunnel the sound energy from the fan includes both inlet and outlet. To obtain the total energy, the above two levels must be added logarithmically.

Sound power levels ref: $10^{-12}$ Watts

| 125 | 250 | 500 | 1K | 2K | 4K |  |
|---|---|---|---|---|---|---|
| 99 | 100 | 98 | 97 | 97 | 90 | dB |

The volume of the tunnel = 43 300 m³.

The reverberation time for the tunnel is assumed to be at least 5 secs based on tests at the Saltash tunnel.

|  | 125 | 250 | 500 | 1K | 2K | 4K | Lin |  |
|---|---|---|---|---|---|---|---|---|
| 10 Log $\left(\frac{4T}{.161V}\right)$ | -25 | -25 | -25 | -25 | -25 | -25 | -25 | dB |
| Total fan SWL | 110 | 111 | 109 | 108 | 108 | 101 | 116 | dB |
| Resultant sound pressure level | 85 | 86 | 84 | 83 | 83 | 76 | 91 | dB |
| Average measured level | 88 | 84 | 86 | 86 | 83 | 75 | 93 | dB |
| NR | 85 | 95 | 91 | 88 | 85 | 83 | 81 |  |

### Saltash Tunnel

Sound pressure levels were recorded at 1.5 metres above the roadway with all fans operating in the form A mode. Twenty readings were taken at 25m intervals down the length of the tunnel with additional readings directly under the fans.

The average of the measured levels in each octave band is :

| 63 | 125 | 250 | 500 | 1K | 2K | 4K | Hz |
|---|---|---|---|---|---|---|---|
| 81 | 88 | 89 | 83 | 82 | 77 | 68 | dB |

The peak levels as measured directly below a group of two fans were only 1 dB higher than the average. The background level was 25 dB lower.

Estimated sound level in tunnel:

The direct level function was ignored since the measured levels on site indicated that at the design measurement point, the direct level contribution was insignificant.

The sound power level measurements recorded in the laboratory on one of the installed fans are:

|  | 125 | 250 | 500 | 1K | 2K | 4K |  |
|---|---|---|---|---|---|---|---|
| Inlet | 94 | 90 | 86 | 86 | 82 | 74 | dB |
| Outlet | 99 | 101 | 89 | 91 | 87 | 80 | dB |
| Total | 100 | 101 | 91 | 92 | 88 | 81 | dB |

For 6 fans installed - total fan sound power level (+ 8dB)

| 125 | 250 | 500 | 1K | 2K | 4K |  |
|---|---|---|---|---|---|---|
| 108 | 109 | 99 | 100 | 96 | 89 | dB |

The volume of the tunnel = 37812 m³.

The reverberation times in each octave were measured at three positions along the length of the tunnel.

|  | 125 | 250 | 500 | 1K | 2K | 4K | Lin |  |
|---|---|---|---|---|---|---|---|---|
| Average rev time-sec | 12.7 | 10.7 | 10.4 | 7.2 | 3.8 | 2.7 |  |  |
| 10 Log 4T | +17 | +16 | +16 | +15 | +12 | +10 |  |  |
| 10 Log $\frac{1}{.161V}$ | -38 | -38 | -38 | -38 | -38 | -38 |  |  |
| Total fan SWL | 108 | 109 | 99 | 100 | 96 | 89 | 114 | dB |
| Resultant Sound Pressure level | 87 | 87 | 77 | 77 | 70 | 61 | 93 | dB |
| Site measurement | 88 | 89 | 83 | 82 | 77 | 68 | 94 | dB |

## 6 DISCUSSION

Although the thrust developed by a jet fan can be tested to a high order of accuracy by the method described, it is not recommended that the values for an apparently geometrically similar fan are scaled from a test. All fan sizes and motor frames should be tested since at the high outlet velocities associated with these fans, a small physical difference can make a significant change in the thrust values.

An accurate measurement of the resultant transfer of the momentum energy to the tunnel airstream is not achieveable with the instrumentation used in the two tunnels described. This is an area which requires further study to establish a reliable method of measuring the pressure rise in a tunnel due to a fan. This would be of considerable help in providing more data on the factors effecting the system performance.

At present the measurement of the airflow in an empty tunnel has been the only effective method of establishing a system performance.

In the two tunnels mentioned, several locations down the length were selected for a full velocity traverse. The position chosen for evaluation on the Saltash was at least 100 metres downstream of a fan and where the velocity distribution was within ±20 per cent of the average in either direction. The variation on Penmaenbach was of the order of +40 per cent and −30 per cent. These results were considered to be of sufficient accuracy to provide a comparison with a theoretical assessment.

In the evaluation of the system resistance, an assumption was made on the tunnel entry factor and also of the friction factor. These values are typical for the majority of tunnels although owing to the tunnel lighting fittings, etc. the friction factor selected might be low.

On the evaluation of the system efficiency a value of 0.59 is derived from the site tests at Penmaenbach and 0.56 at Saltash with the fan operating form A and form B respectively. Form A tests at Saltash indicated a value of 0.88. No tests were carried out at Penmaenbach with fans running form B.

Some work has been done to establish by laboratory tests the friction losses on a wall caused by a jet flow (Ref 1,2). From this and other data (Ref 3) suggest that the Penmaenbach has a loss of about 10 per cent of the fan thrust whereas Saltash has a loss of about 5 per cent. On the latter installation the fans are located 2.5m from the ceiling with the parallel fans at 1.6m spacing on their axes.

As mentioned in this paper, there are other parameters which contribute to the relatively low overall efficiency. Values of between 0.5 to 0.9 are to be expected (Ref 4, 5).

The distance from the discharge outlet of the fan to the inlet of the next is important. Ideally the second fan should be located where the velocity profile across the tunnel is uniform. This distance increases with tunnel velocity and it is suggested (Ref 6) that at 4 m/s the distance with 610mm jet fans is 140m. Also reference is made to the distance of the jet fan from the tunnel entrance, stating that the ventilation efficiency is approximately 80 per cent at 100m.

No explanation can be found for the apparent factor of 0.88 for the fans operating in the Form A direction at Saltash. Both tunnel portals and the topography are similar. The laboratory tests on one of the fans gave equal thrust in either direction.

The overall sound pressure levels measured in both tunnels were in close agreement with the calculated levels (linear), within 2dB. However, the Saltash measurements deviated by up to 7dB in the octave bands above 500Hz. This could be due to a focussing effect since the readings were taken in the centre of the roadway with a circular ceiling above.

7    CONCLUSIONS

Although this investigation has shown low system efficiencies for the jet fan with an empty tunnel, this does not relate to the traffic flow condition where the design requirement is for a lower velocity where traffic drag is a major component. The aerodynamic losses as discussed will reduce quadratically with the velocity.

The noise level predicted in a tunnel can be established with reasonable certainty from sound power data measured in the laboratory by the test method described.

The thrust from a jet fan can also be accurately measured but the resultant system performance is not so readily predicted. More work needs to be done to find out methods of improving jet fan system efficiency. This is now becoming an important issue since longer tunnels are now being ventilated by this method with the consequence of fans having to be sited closer together.

REFERENCES

(1)  ROHNE, E.  The friction losses on walls caused by jet flows of booster fans.  3rd International Symposium on the Aerodynamics and Ventilation of Vehicle Tunnels (Sheffield, 19-21 March 1979) Cranfield, UK, BHRA, Paper C1 pp.57-70

(2)  ROHNE, E.  The influence of axis distance of two parallel jet flows on the friction losses on walls.  5th International Symposium on the Aerodynamics and Ventilation of Vehicle Tunnels (Lille, France, 20-22 May 1985) Cranfield, UK, BHRA, Paper J2, pp.543-552

(3)  REALE, F.  Fundamentals and Applications of Induction Ventilation Systems of Vehicle Tunnels.  1st International Symposium on the Aerodynamics and Ventilation of Vehicle Tunnels (Canterbury, April 1973) Cranfield, UK, Paper H1, pp.1-16

(4)  GIUBILO, M., LACQUANITI, V.  Uncertainties in the design of longitudinal ventilation systems for bidirectional road tunnels.  4th International Symposium on the Aerodynamics and Ventilation of Vehicle Tunnels (York, March 1982) Cranfield, UK, Paper K3, pp.461-482

(5)  P.I.A.R.C. TECHNICAL COMMITTEE: Report on Road Tunnels.  XVIth World Road Congress (Vienna, Austria, 16-21 September 1979) pp.50-51

(6)  BABA, T.  Characteristics of longitudinal ventilation system using normal size jet fans.  3rd International Symposium on the Aerodynamics and Ventilation of Vehicle Tunnels (Sheffield, March 1979) Cranfield, UK.  Paper C2 pp.71-92

(7)  SHARLAND, I.  Woods Practical Guide to Noise Control. Woods of Colchester Ltd

ACKNOWLEDGEMENTS

The authors acknowledge permission from The Department of Transport, Exeter and The Highway Directorate, Welsh Office, to publish this paper.

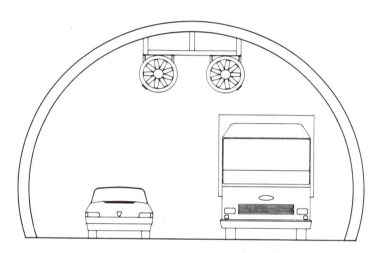

Fig 1a    Jet fan location in circular tunnel

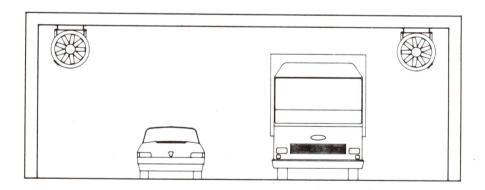

Fig 1b    Jet fan location in rectangular tunnel

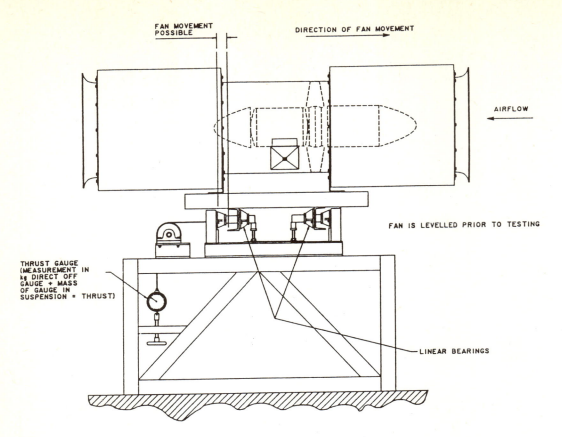

Fig 2     Thrust measurement rig

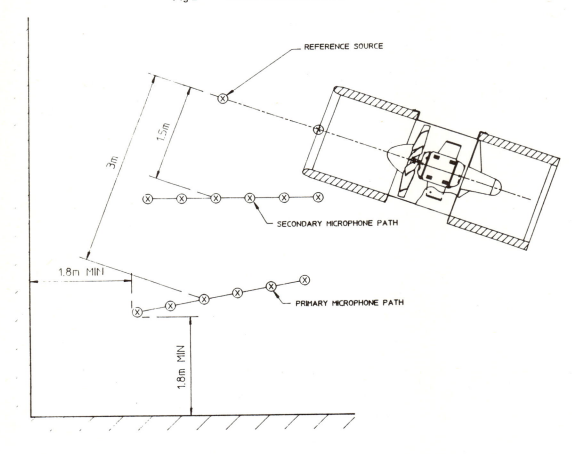

ALL SURFACES HARD FINISH

Fig 3     Noise measurement — semi-reverberant test method

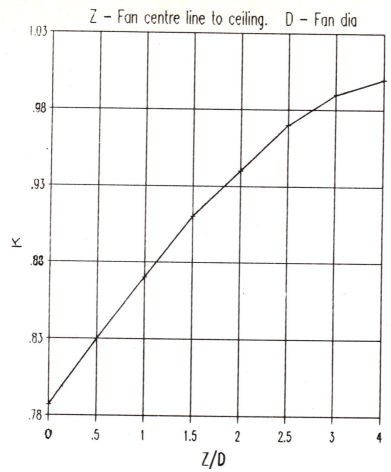

Fig 4    Fan proximity to ceiling factor

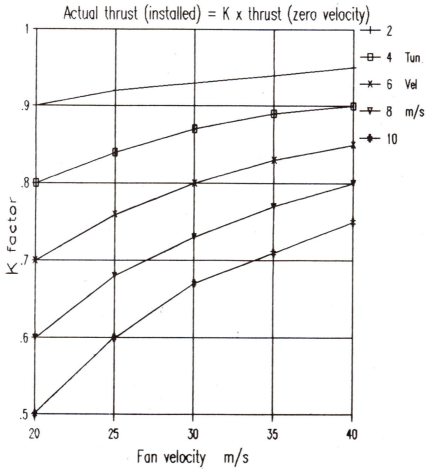

Fig 5    Effect of tunnel velocity on fan thrust

# Fans for Class 91 locomotives

**N P HODGSON**, BSc, AMIEE
GEC Alsthom, Transportation Projects Limited, Manchester

SYNOPSIS The Class 91 locomotive uses fans for cooling smoothing chokes, traction motors, oil radiators and brake resistors. The fans have to deal with the onerous conditions associated with a
railway traction environment, while keeping size and weight to a minimum.

## 1. INTRODUCTION

The British Rail Class 91 locomotive is a 25 kV
locomotive with a rated power of 4.54 MW and a
design speed of 225 km/h. The locomotive will
haul up to 11 MkIV coaches and a driving van
trailer from London to Edinburgh. In addition,
sleeper trains of up to 750 tonnes may be hauled
on the West Coast Main Line including the ardu-
ous ascents of Shap and Beattock.

The contract was placed by British Railways
Board in February 1986 and the first locomotive
was handed over on 14th February 1988. The locomotive utilises fans to cool traction motors,
chokes, transformer and converter radiators,
braking resistors and air conditioning. This
paper will describe the factors affecting fan
choice (with the exception of air conditioning)
and give some examples of the problems which the
fans encounter, when operating in a railway
environment.

## 2. THE LOCOMOTIVE AIR MANAGEMENT SYSTEM

Figure 2 shows schematically the air management
system.

### 2.1 Traction motor blower

The Class 91 is a four axle locomotive, with two
axles in each bogie and all axles individually
powered.

Figure 3 shows the power circuit for one
bogie.

The d.c. traction motors are powered via
smoothing chokes and thyristors from the locomotive transformer. Each of the four traction
motor blowers provides cooling air for an
individual traction motor and its associated
choke.

Each fan is required to supply $1.65m^3$/s at
2.29 kPa. With such a large head to flow rate
requirement the choice of fan was confined to a
centrifugal type. Limitations on space
necessitated the use of a forward curved
impeller.

### 2.2 Cooler group fans

Separate radiators are used for the oil circuits
which cool motor control thyristors and the
locomotive transformer. The two radiators are
mounted on the transformer, and are cooled by
two cooler group fans.

Each fan is required to supply $3.5m^3$/s at
1.24 kPa. With such a large flow rate to head
requirement the choice of fan was restricted to
an axial type.

Of the $3.5m^3$/s of air which is delivered,
$2.3m^3$/s is passed over the radiators, the re-
mainder is bled off into electrical equipment
cubicles to provide cooling. After passing
through the cubicles, the air enters the locomotive body and then leaks out to atmosphere.

### 2.3 Rheostatic braking resistor

The traction motors have their separately excited fields reversed during braking, thereby
turning the motors into generators. The generated energy is then dissipated into two fan
cooled resistors; one per bogie.

With a requirement for a large flow rate
($9 m^3$/s) at a low head (1 kPa) an axial fan was
chosen.

The fan draws its air from the locomotive
underframe, forces it over the resistor and then
exhausts it to atmosphere through the roof.

As air passes over the hot resistors its
density reduces. This has a significant effect
on the system head. The impeller is, therefore,
chosen to suit the conditions which occur when
the resistor is hot.

## 3. FAN MOTORS

### 3.1 Traction motor blowers and cooler group fans

The same type of motor is used for traction
motor blowers and cooler group fans. It is a
7.5 kW, 240 V, single phase a.c., capacitor
start and run machine. This motor rating represents the limit at which single phase a.c.
machines are commercially available.

The supply for the motors, for simplicity,
is obtained directly from a tertiary winding of
the locomotive transformer. Extra power requirements would have necessitated the use of either
inverter fed three phase a.c. machines or d.c.
machines.

Operating requirements for the locomotive
include a temperature range of -25°C to +35°C
and a supply voltage variation of 62 per cent to
110 per cent of nominal.

The machine is rated to cope with these extremes, the worst case being minimum temperature (where air density has risen by 20 per cent from its value at 20°C) and minimum voltage.

## 3.2 Rheostatic brake fan motor

The power supply for the d.c. rheostatic brake fan motor is obtained by tapping it directly across part of the resistor which it cools. This obviates the need for a control circuit for the fan motor. It also increases locomotive efficiency, by using power which would otherwise be dissipated as heat in the resistors.

The motor rating takes into consideration many factors, including inlet air density variation and the change in head experienced by the fan as the resistor above it heats to its 600°C operating temperature.

Each motor is rated at 24 kW, but recognising that long periods of braking cannot occur on British Railways, only for half an hour.

## 4. LOCOMOTIVE TESTING AND SERVICE REQUIREMENTS

## 4.1 Fan installation

All traction equipment is required to survive in the harsh environment of a railway vehicle. It must, therefore, be securely mounted to the vehicle body. To ensure that equipment satisfies these requirements the following load cases are applicable :

i) Proof load cases

| | |
|---|---|
| Vertical up | 1.0 g |
| Vertical down | 1.5 g |
| Lateral | 1.1 g |
| Longitudinal | 3.0 g |

ii) Fatigue load cases

| | |
|---|---|
| Vertical | 1 ±0.3 g |
| Lateral | ±0.3 g |
| Longitudinal | ±0.2 g |

In addition the fans must not generate any natural frequencies that would interact with other items causing resonances. The fans must also be able to tolerate vibration frequencies of up to 30 Hz, emanating from the natural characteristics of railway vehicles.

## 4.2 Airflow testing

Part of the testing programme for the locomotive, involved subjecting the major items of electrical equipment to pseudo-service conditions on a simulator, several months before the first locomotive was built.

During these tests, the airflows associated with the traction motor blowers and cooler group fans were checked (the rheostatic brake resistor's air flows were type tested separately). Adjustments to guide vanes in the cooler group system were necessary to obtain the correct air flows through the equipment cubicles.

Airflow tests on the locomotive are confined to testing while the locomotive is static. Dynamic airflow tests being impractical. The dynamic performance of the fans is monitored by ensuring that the temperature rises of the fan cooled equipment is within specification.

It was noted that the airflows were higher on the locomotive than those obtained during the BS 848 standardised airway tests. It would appear that the cause of this discrepancy was the louvre, because when the louvre was removed agreement better than three per cent was achieved.

## 4.3 Washing plant

A typical day can require a locomotive to cover more than 1000 km, consequently becoming very dirty.

The locomotive is the flagship of British Railway's InterCity fleet, therefore to present a good public image daily washing is a necessity.

It is important, that, as the locomotive passes through a washing plant water is not blown into the traction motors by the traction motor blowers. Water in the traction motors can cause flashover.

British Rail's washing plants pose two onerous problems.

Firstly, they use rotating flat polypropylene strips together with a cleaning solution to remove accumulated dirt.

Secondly they have a final rinsing process which through jets deliver 7 l/s of water per side at a pressure of 140 kPa.

To have a bodyshell as light and rigid as possible the size of the bodyside air inlets were kept to a minimum. The original design for the air inlet to the traction motor had a single louvre, but with louvre velocities as high as 10m/s moisture separating capabilities were found to be inadequate.

Several solutions to this problem were tested. The one which was found to perform best comprised the same bodyside louvre, with the addition of a plenum chamber and secondary louvre prior to smoothing choke and traction motor. Water, which is separated in the plenum chamber and secondary louvre, is drained off via a manometric trap.

## 4.4 Snow

British Rail's experience when operating High Speed Trains (HSTs) in snow, had shown snow accumulations on the bodysides at the rear of trains. This is not due to falling snow, but snow already lying on the ground being picked up by the air turbulence associated with the train.

To see how the Class 91 fared under similar conditions, the traction motor ventilation system was mocked up and tested in a climatic chamber.

No British data is available for snow particle size and mass flux (density), at varying heights above the ground. Tests were therefore based upon a Canadian Military specification, MIL210C, which contains such data.

Snow particles varying in size from a nominal 30 μm to a nominal 120 μm were produced by mixing compressed air and water in a snow gun. Problems with the freezing of the gun precluded testing at temperatures lower than -15°C. Mass fluxes as high as 6 g/m²/sec were used.

Under all circumstances simulated during the tests, the system was found to operate satisfactorily. The louvres did not block and the quantity of snow passing through to the traction motors was not of any consequence.

### 4.5 Noise

British Rail specifications dictate acceptable levels of external and internal noise both at speed and at rest. Calculations were undertaken at the design stage to verify the acceptability of the noise levels of proposed fans, recent type testing has shown these calculations to be valid.

To meet stationary noise levels it is necessary to switch off the traction motor blowers and cooler group fans. However, equipment within the cubicles still requires ventilation. Therefore two small axial fans are provided for this purpose to extract heat from the cubicles to the locomotive body.

### 5. EXPERIENCE TO DATE

Type testing of the locomotive propulsion equipment has been completed satisfactorily, with the temperature rises of smoothing chokes, traction motors, oil radiators and brake resistors all proving acceptable.

Subsequent to this testing the first locomotive entered passenger service in March 1989, seven months ahead of contract date.

### 6. ACKNOWLEDGEMENT

The author would like to thank his colleagues within GEC ALSTHOM and Woods of Colchester for help in the preparation of this paper.

He also thanks British Railways Board and GEC ALSTHOM for permission to present the paper

Fig 1    Class 91 locomotive

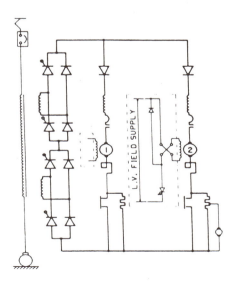

Fig 3    Bogie power schematic

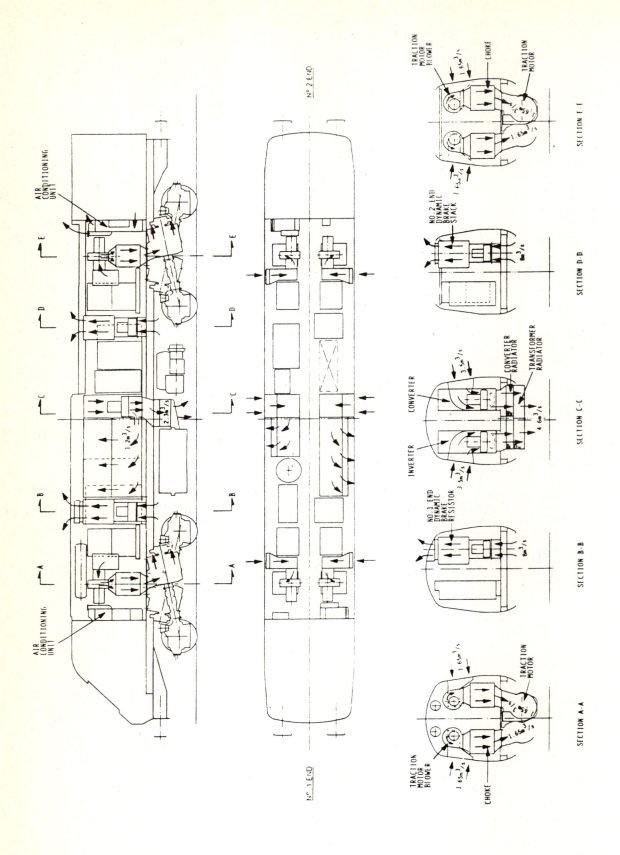

Fig 2    Locomotive air management system

Fig 4    Cooler group fan and equipment cubicle

Fig 5    Traction motor blower with choke, plenum chamber, equipment
cubicle and rheostatic brake stack

# C401/033

# Some problems with the design and installation of engine driven automotive cooling fans

E H FISHER, BSc, PhD, CEng, MIMechE
Department of Mechanical, Materials and Manufacturing Engineering, University of Newcastle Upon Tyne

SYNOPSIS. Modern trends in car design place increased demands on the coolant system designer.
Features of the coolant duct which could be improved by more careful aerodynamic design are
identified. The limitations of viscous coupled engine driven fans are also reviewed. By reference
to a particular design study, installation effects such as tip clearance and axial separation from
the radiator matrix are shown to predominate over basic fan design.

NOTATION

| | |
|---|---|
| $c_x$ | axial velocity component at distance x from the rotor |
| $c_{x_1}, c_{x_2}$ | axial velocity components far upstream and downstream |
| $c_{\theta 2}$ | swirl velocity component downstream of the rotor |
| d | fan diameter |
| $\Delta H_o$ | stagnation head rise |
| K,k | coefficients |
| L | sound power level, dB |
| r | radius |
| $r_h, r_t$ | hub and tip radii |
| SPR | sound power ratio (= sound power/air power) |
| x | axial distance with origin at the fan |
| $\Omega$ | angular velocity |

## 1. INTRODUCTION

Modern trends in car design demand smaller frontal air intakes and produce increasingly restricted engine compartments. The result is that medium and large engined cars frequently have difficulty meeting cooling test conditions. The problem may emerge as failure to cool the engine, excessive fan noise, or excessive fan power consumption.

The principal features of a modern cooling duct are described in Section 2 of the paper, and those facets which should be optimised within packaging constraints are identified.

Low speed cooling requirements are particularly dependent on fan performance. The viscous coupled engine driven fans which are still common on larger cars are far from ideal for this duty, as discussed in Section 3.

Space limitations impose severe restrictions on the fan designer. The general constraints are reviewed in Section 4, while a description of a specific application is presented in the final section. It is clear that axial position of the fan relative to the radiator and radial clearance between the fan and cowl have first order effects on performance.

## 2. THE COOLANT AIR DUCT

Modern cooling systems are required to perform demanding duties within severe space constraints. A typical 150 H.P. (112 kw) engine will shed some 85 kw of heat to coolant, requiring an airflow of over 2 $m^3$/s. For vehicles in the Grand Touring class having engines of some 300 H.P.(225 kw) these figures will approximately double.

The principal features of a 3-litre V6 installation are shown in Figure 1; the dimensions given below are non-dimensional with respect to the fan diameter, d. Air enters through the moulded plastic intake, which terminates abruptly. After expansion, cooling air passes through the radiator and enters a cowl which is fixed to the radiator. This extends some 0.13 d downstream and surrounds the cooling fan with a tip clearance of about 0.05 d to allow for movement between engine and body mounted components. Upstream and downstream clearance from the fan to the radiator matrix and engine block are approximately 0.1 d and 0.2 d respectively.

It is significant that most of the system resistance occurs away from the radiator, principally with expansion losses upstream and undue restriction in the engine compartment downstream. Even worse situations exist in the Grand Touring class where the engine compartment can impose several times more resistance than the radiator matrix. It is indicative of the changes which have taken place in recent years that earlier work by Hay and Taylor (1) did not consider upstream losses at all, and adopted a downstream geometry which restricted flow by less than 2% of the radiator matrix resistance.

It is clear that greater attention to the aerodynamic design of cooling ducts would reduce cooling problems. Radical new approaches have been suggested by several authors. Buchheim et al (2) claimed a substantial reduction in vehicle drag when air was discharged through a bonnet louvre. Although such systems are to be found in a number of sports/racing cars, they seem unlikely to find widespread acceptance in the market. Paish and Stapleford (3) have devised a ducted system which promises reductions of 2-4 times in radiator area.

Problems of accommodating the ducting within a vehicle and with adequate fan assistance for low speed cooling remain barriers to its implementation.

It seems likely that refinement of current systems will be necessary for at least some years to come. Wide angle diffusers coupled with splitter vanes will be used increasingly to reduce expansion losses. More realistic outlet ducts discharging to regions of low pressure will further improve the position. These will have to be developed in every case within the packaging constraints of each particular vehicle.

## 3. COOLING FAN DUTY

Although the purpose of the cooling system is to cool the engine in all conditions, manufacturers evaluate performance at a set number of test points. These differ in detail from manufacturer to manufacturer, but will invariably include:

(a) an idle cooling requirement
(b) one or more low speed conditions at full throttle, or a specified vehicle weight, or simulating a specified gradient
(c) one or more high speed conditions in the range 90-100% of maximum speed.

The airflows for a 3-litre V6 engine cooled by a viscous coupled fan are shown in Figure 2. The results were obtained by integration of velocity measurements across the radiator face recorded in road tests. The uncertainties of this procedure account for the generous tolerance bands placed on results.

In the speed range considered, it would appear that in top gear the cooling fan increases flow by only about 10% above ram air values. Tests were restricted to legal road speeds below 112 k.p.h., but the disengaged fan results would be expected to extrapolate linearly to the maximum vehicle speed of approximately 200 k.p.h. In this condition the airflow would be 2.25 $m^3/s$, which was as expected more than adequate to satisfy high speed cooling test points. The results for the engaged fan cannot be extrapolated linearly to 200 k.p.h. for the reason shown in Figure 3. Above a coupling input speed of about 3000 RPM, corresponding roughly to a vehicle speed of 112 k.p.h., the fan speed no longer increases linearly with engine speed, but approaches its maximum value. The effect of fan assistance at maximum speed would thus be even less than the 10% evident at lower speeds. Indeed, Hay and Taylor (1) have shown that, for conditions with a strong ram air element, a driven fan may actually reduce airflow below that achieved by ram air alone. All the evidence indicates that the fan does not have a significant effect on high speed cooling.

A different situation exists at the low speed and idle conditions, where the test vehicle was known to be more marginal on cooling test. At 30 k.p.h. the fan assisted flow is some three times greater than the ram air value, indicating that the fan makes up almost 90% of the head loss through the system.

It is unfortunate that this vital duty must be achieved at only 0.6 of fan peak speed (Figure 3). A fan matched to this duty would consume over twelve times as much power if engaged at maximum engine speed. The effect on noise is perhaps even more significant in a modern car. As sound power, L, is linked to fan speed, $\Omega$, by

$$L_2 - L_1 = K \log\left(\frac{\Omega_2}{\Omega_1}\right) \quad dB \qquad (1)$$

where $55 < K < 60$

the maximum noise could be 13 dB above the level required to satisfy the cooling condition.

It would appear that electrical fans matched to the cooling condition offer a better solution in the vast majority of cases. The fan power required will be less than 250 watts for most 3-litre vehicles, which is within the capacity of modern 12-volt systems.

## 4. SOME PROBLEMS OF COOLING FAN DESIGN

The objective of any cooling fan design must be to satisfy the system head/flow requirements with minimum noise levels.

Rig tests show that production fans are unlikely to exceed about 60% efficiency. Preliminary analysis indicates that the greatest source of loss is downstream kinetic energy dissipated from the rotor only designs.

It would appear that adding a stator row would improve the position, giving improved efficiency and reduced speed with consequent sound reductions. It is clear from the dimensions of Figure 1, however, that this cannot be accommodated in present designs. The introduction of a second blade row may also introduce discrete blade passing frequencies in addition to broad band noise.

An even more significant restriction is imposed on blade design by the dimensions shown in Figure 1. Intuitively a low hub/tip ratio design would be chosen when the fan is in such close proximity to the radiator. Unfortunately, the variation in head rise at different blade sections which such designs inevitably produce gives rise to axial velocity gradients both upstream and downstream of the rotor. Following the actuator disc approach of Horlock (4), this influence extends upstream and downstream in accordance with the relationships:

$$c_x = c_{x_1} + \left(\frac{c_{x_2} - c_{x_1}}{2}\right) \exp(kx/(r_t - r_h)) \text{ where } x<0, \text{ upstream}$$

$$c_x = c_{x_2} - \left(\frac{c_{x_2} - c_{x_1}}{2}\right) \exp(-kx/(r_t - r_h)) \text{ where } x>0, \text{ downstream}$$

As $k \approx \pi$ axial velocities are affected significantly for at least one blade height either side of the rotor. It is clear from Figure 1 that this would result in serious interaction with the radiator matrix and significant interaction with the engine block. The position is exacerbated by the fact that an engine mounted fan will frequently be offset from the radiator matrix. It would appear that until CFD programs capable of modelling the entire radiator/cowl/fan regime are available, the designer is restricted to free vortex designs, which do not suffer such axial perturbations.

Reducing fan speed appears superficially to be an attractive means of reducing noise. It follows from equation (1) that a 20% speed reduction would decrease noise by over 5 dB for a given fan. Unfortunately the fan duty would also reduce substantially in accordance with the affinity laws.

No analytical means exists for comparison of the acoustic performance of different fans, but Deeprose (5) has suggested the use of a Sound Power Ratio

$$SPR = \frac{Sound\ power\ (watts)}{Air\ power\ (watts)} = \frac{4(\Omega r_t)^2}{10^{16}}$$

where the expression involving $\Omega$ is based on empirical correlation. If this relationship is applied to two axial fans producing the same duty at speeds differing by 20%, the slower fan might be expected to be some 2 dB quieter. It is also likely that improvements in aerodynamic design would yield further noise reductions.

Reducing fan speed for a given duty is, however, likely to reduce efficiency. This follows from the Euler equation

$$\Delta H_o = \frac{(\Omega r)}{g} C_{\theta 2}$$

which indicates that lower rotational speed, $\Omega$, leads to higher downstream swirl losses. Efficiency will therefore be reduced, unless this effect can be offset by improvements in aerodynamic design.

## 5. A COMPARISON OF TWO FAN DESIGNS

From the general considerations discussed above, it was decided to design a prototype fan matching the duty of the production unit, but at a reduced speed of 1250 RPM compared with 1500 RPM.

The flow requirement was known from vehicle tests of the type shown in Figure 2. The performance of the production fan, a nine bladed rotor of hub/tip ratio approximately 0.45:1, was established on a test rig which approximated to a BS 848 Type C installation.

Results of flow vs. head are presented in Figure 4. The most striking observation is that the production fan is clearly stalled in the flow range anticipated. It seemed that this might be a significant source of noise: rapid increases in noise from fans operating below their stall flow have been reported by Deeprose (5) in centrifugals and by McEwen and Margetts (6) in mixed-flow machines.

However, preliminary sound measurements taken with a simple sound meter some 5 cm downstream of the rotor failed to show any significant variation with flow for either the production or prototype fan. Further work is needed to establish the precise reason for this: the results could well have been

influenced by either the axial position of the sound probe or the absence of an effective wind shield around it.

It could also be significant that the boundary layer displacement thickness on the test rig ducting exceeded the tip clearance, which could lead to flow disturbance in the tip region at all flows. As noise generation varies with a high power of blade speed, this tip region might be expected to dominate noise generation. If such an explanation is valid, it makes the extrapolation of rig sound measurements to the vehicle extremely difficult.

A prototype fan was designed and manufactured to match or exceed the production fan performance. The twin constraints of free vortex flow and reduced speed limited the hub/tip ratio to 0.7. Axial length constraints also necessitated the use of 21 blades. As shown in Figure 4, the test rig performance satisfied the design objectives. Head exceeds that of the production fan in the critical flow range, and stall has been moved to the lowest flow expected. Efficiencies for the two fans are similar, with the inherent problems of lower speed being offset by greater attention to aerodynamic design.

Vehicle tests failed to confirm the superior performance of the prototype fan, and necessitated a further series of tests with the fans mounted in the radiator/cowl assembly. The results of these are summarised in Figures 5 a-b. It is clear that the prototype fan is only inducing some 70% of the flow obtained with the production unit; this flow is reasonably independent of axial distance from the radiator matrix in the range of interest, 0.06 d – 0.13 d.

A fundamental change was observed when the tip clearance of approximately 0.05 d was sealed. With the fan in the optimum position, some 0.13 d from the radiator matrix, the production fan flow increased over 20% and the prototype fan flow by over 70%. Even the lower improvement for the production fan implies that fan speed could be reduced by 20%, yielding a power saving of up to 50% and a noise saving of over 4.5 dB.

In addition, both fans were now sensitive to axial position. Comparison with the design data given by Tenkel (7) shows that the production fan is some nine times more sensitive to axial position than was predicted from his data. It is also clear that the high hub/tip ratio prototype fan is influenced some 2.5 times more than the low hub/tip ratio production unit. It is concluded that the influence of the radiator matrix in close proximity to cooling fans is a function both of the hub/tip ratio of the fan and of the installation being tested.

## 6. CONCLUSIONS

The difficulties experienced in cooling modern cars would be reduced if greater attention were paid to the coolant air duct design: expansion losses upstream of the radiator and duct design downstream are prime areas for improvement.

Viscous coupled engine driven fans are far from ideal for current requirements. In a particular study, it was shown that the viscous fan could absorb twelve times more power and generate 13 dB more noise than is necessary.

For many cases, installation effects predominate over basic fan design. For a particular study, both tip clearance and axial separation from the radiator were seen to have first order effects.

Effective sealing of the tip clearance might halve the power requirement and reduce noise over 4.5 dB. In contrast, a fan redesign produced little improvement in efficiency and a noise saving of not more than 2 dB.

In summary, installation effects are of prime importance in automotive cooling fan systems.

REFERENCES

(1) HAY,N. and TAYLOR,S.R.G. The effects of vehicle cooling system geometry on fan performance. I.Mech.E.Conf. 'Fan Technology and Practice', April 1972, pp.176-192.

(2) BUCHHEIM,R., DEUTENBACH,K-R. and LUCKOFF,H-J. Necessity and Premises for reducing the Aerodynamic Drag of Future Passenger Cars. SAE Paper 810185 (1981).

(3) PAISH,M.G. and STAPLEFORD,W.R. A rational approach to the aerodynamics of engine cooling system design. Proc. I.Mech.E. Vol.83,Pt.2A,No.3, (1968-9).

(4) HORLOCK,J.H. Actuator Disc Theory. McGraw-Hill, New York, 1978.

(5) DEEPROSE,W.M. and BROOKS,J.M. Effect of Scale on Fan Noise Generation of Backward-Curved Centrifugal Fans. I.Mech.E.Conf. 'Fan Technology and Practice', April 1972, pp.137-153.

(6) McEWEN,D. and MARGETTS,E. The effect of rotor blade variations on the performance and noise of a mixed-flow fan. NEL Fluid Mechs.Silver Jubilee Conf.East Kilbride, Nov.1979.

(7) TENKEL,F.G. Computer Simulation of Automotive Cooling Systems. SAE Paper No.740087, New York, 1974.

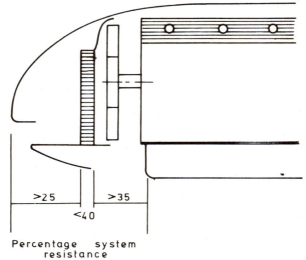

Percentage system resistance

Fig 1    Approximate dimensions of V6 cooling system

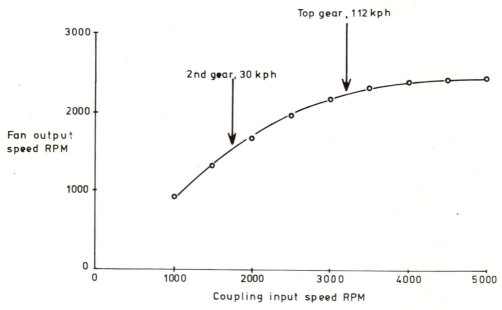

Fig 3    Viscous coupling speed characteristics

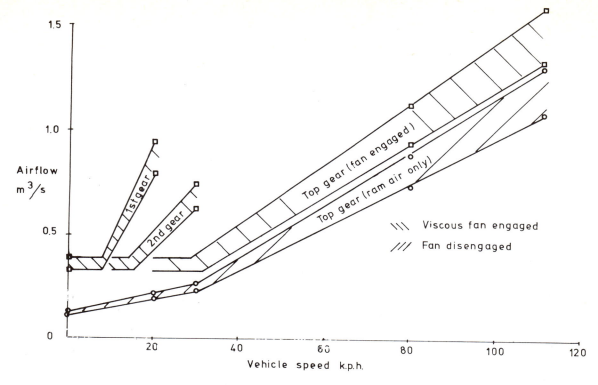

Fig 2     Coolant airflows for 3 litre V6 engine

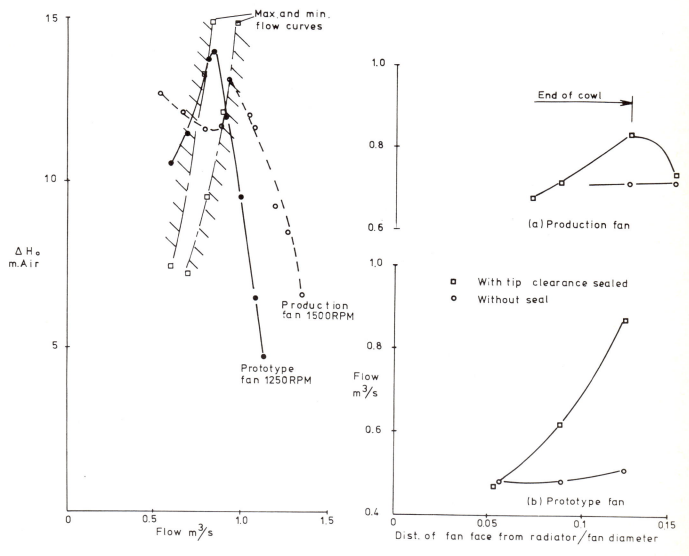

Fig 4     Fan test rig flow/head characteristics

Fig 5     Radiator/cowl flow tests